Engineering Materials

This series provides topical information on innovative, structural and functional materials and composites with applications in optical, electrical, mechanical, civil, aeronautical, medical, bio- and nano-engineering. The individual volumes are complete, comprehensive monographs covering the structure, properties, manufacturing process and applications of these materials. This multidisciplinary series is devoted to professionals, students and all those interested in the latest developments in the Materials Science field, that look for a carefully selected collection of high quality review articles on their respective field of expertise.

Indexed at Compendex (2021) and Scopus (2022)

Elena Lysenko · Alexander Rogachev ·
Olga Galtseva
Editors

Emerging Trends in Materials Research and Manufacturing Processes

Springer

Editors
Elena Lysenko
Tomsk Polytechnic University
Tomsk, Russia

Alexander Rogachev
Research Institute of Physics and Chemistry
Francisk Skorina Gomel State University
Gomel, Belarus

Olga Galtseva
School of Non-Destructive Testing
Tomsk Polytechnic University
Tomsk, Russia

ISSN 1612-1317 ISSN 1868-1212 (electronic)
Engineering Materials
ISBN 978-3-031-38963-4 ISBN 978-3-031-38964-1 (eBook)
https://doi.org/10.1007/978-3-031-38964-1

This Springer imprint is published by the registered company Springer Nature Switzerland AG
The registered company address is: Gewerbestrasse 11, 6330 Cham, Switzerland

Preface

The technological breakthrough at this stage is associated with digital transformation. It is assumed that innovations from different industries interact in a complex way. At the same time, fundamental research and its industrial implementation underlie the developed products and technologies and are aimed at improving modern technological processes and achievements. However, digital transformation not only opens up new opportunities but also creates additional risks. The book presents the latest developments and new directions in advanced control systems, as well as new theoretical discoveries, industrial applications, and case studies of complex engineering systems and materials science. The authors thank the Springer Nature team for cooperation.

Keywords

Technology, Materials, Monitoring, Modeling, Measurement, Complex Engineering Systems, Control Systems, Quality, Processing, Testing, Thermal Analysis and Stabilization, Sensors, Electronics, Modeling

Tomsk, Russia — Elena Lysenko
Gomel, Belarus — Alexander Rogachev
Tomsk, Russia — Olga Galtseva

Contents

The Influence of Thermomagnetometric Measurement Conditions on the Recorded Curie Temperature of Cobalt-Zinc Ferrite

Evgeniy Nikolaev, Elena Lysenko, Anatoly Surzhikov, and Sergey Bobuyok

Abstract The influence of the measurement atmosphere, heating and cooling rate on the Curie temperature of $Co_{0.5}Zn_{0.5}Fe_2O_4$ cobalt-zinc ferrite was studied by thermogravimetric analysis in a magnetic field. This technique allowed determining the temperature of ferrimagnet-paramagnet transition at Curie point of magnetic materials. The method of solid-phase synthesis was used to produce the Co–Zn ferrite. Curie temperature control of the ferrite was carried out by using a Netzsch STA 449C Jupiter thermal analyzer. Measurements were performed in the temperature range of 45–400 °C with heating (cooling) rates of 10, 20, and 50 K/min. The cell of analyzer was purged with air or nitrogen. According to thermomagnetometric analysis, the Curie temperature of Co–Zn ferrite are in the range of 171−188 °C, depending on the experimental conditions. It was shown, that the Curie temperature weakly depends on both the heating and cooling rate during thermogravimetric analysis in a magnetic field. However, this parameter depends on the measurement mode.

1 Introduction

Currently, magnetic materials based on multicomponent ferrites [1–3] are the key components to design and manufacture modern magnetic and radio engineering devices [4–8]. A wide class of soft magnetic ferrite materials includes cobalt-zinc ferrospinels with the chemical formula $Co_{(1-x)}Zn_xFe_2O_4$ (Co–Zn). These materials are commonly used in various electrical devices. In addition, highly dispersed

E. Nikolaev (✉) · E. Lysenko · A. Surzhikov · S. Bobuyok
National Research Tomsk Polytechnic University, 30 Lenin Avenue, Tomsk 634050, Russia
e-mail: nikolaev0712@tpu.ru

E. Lysenko
e-mail: lysenkoen@tpu.ru

A. Surzhikov
e-mail: surzhikov@tpu.ru

S. Bobuyok
e-mail: sab45@tpu.ru

E. Lysenko et al. (eds.), *Emerging Trends in Materials Research and Manufacturing Processes*, Engineering Materials, https://doi.org/10.1007/978-3-031-38964-1_1

powders of Co–Zn ferrites are widely used in hyperthermia [9, 10], drug transportation [11, 12], magnetic storage [13] and magnetic resonance imaging [14]. It should be noted that the cobalt monoferrite ($CoFe_2O_4$) substituted with Zn ions can exhibit a deformed spinel structure. This depends on the cobalt-zinc ratio in the mixture of initial reagents. The position of cations and their distribution in substituted ferrites can be controlled, which allows changing their structural, electrical and magnetic properties. The Curie temperature is one of the most important parameters characterizing the magnetic properties of ferrite materials of complex composition.

There are several methods for determining the Curie temperature of materials [15–17]. Most of them come down to heating a ferromagnetic sample and determining the temperature, which corresponds to a sharp decrease in magnetization. This process occurs in a constant magnetic field. The variety of methods consists in ways to measure magnetization or to observe its behavior.

In our previous works [18–20], a method (thermogravimetric analysis in magnetic field) for determining the Curie temperature of substituted ferrites was studied. Thermogravimetric analysis (TG) is a method of thermal analysis based on recording the change in the mass of samples depending on heating (cooling) at a given rate, as well as on the composition of the gaseous atmosphere. Moreover, TG analysis is widely used to study the kinetics of physical and chemical processes, as well as to study the nonstoichiometry of materials [21–23]. As a result, curves of the change in the mass of the sample as a function of temperature or time (TG-curve) or the rate of change in mass (differential thermogravimetric (DTG) curve) are obtained. The DTG curve allows determining the time or temperature, which corresponding to the Curie point of the samples. This technique of determining the Curie temperature of ferrites, which based on TG analysis in magnetic field, is called the thermomagnetometric analysis.

In this work, the influence of the heating (cooling) rate and measurement atmosphere during thermomagnetometric measurements on the values of the Curie temperature of nickel-zinc ferrite was investigated.

2 Research Method

Cobalt-zinc ($Co_{0.5}Zn_{0.5}Fe_2O_4$) ferrite was used as the test material. This ferrite was synthesized by solid-state synthesis. Commercial oxides Fe_2O_3, ZnO (analytically pure) and Co_3O_4 (chemically pure) were used as initial reagents. The oxides were pre-dried in a laboratory furnace at 200 °C for 120 min and then mixed in an agate mortar in the required proportions according to the reaction equation:

$$0.5Co_3O_4 + 3Fe_2O_3 + 1.5ZnO \rightarrow 3Co_{0.5}Zn_{0.5}Fe_2O4 + 0.25O_2$$

After that, the mixture of initial reagents Fe_2O_3–ZnO–Co_3O_4 was mechanically activated 60 min in air with using a Retsch Emax high energy ball mill. The rotation speed of the grinding bowls was 2000 rpm. Mechanical activation was performed using stainless steel grinding bowls and balls. The diameter of grinding balls was 5 mm. The ball-to-powder weight ratio amounted to 1:10.

Prior to synthesis, samples were compacted in the form of pellets with diameter of 15 mm. Pellets were made by single action cold pressing. Cobalt-zinc ferrite was synthesized in the laboratory furnace at 900 °C for 4 h. During synthesis, the temperature was controlled using a platinum–platinum–rhodium thermocouple.

The formed ferrite phase was studied by X-ray phase analysis on an ARLX'TRA diffractometer. An X-ray tube with a copper anode was used as a source of X-ray radiation, and a semiconductor detector with a Peltier cooler (Peltier detector) with a resolution of 250 eV was used as a receiver of inelastic scattering. The diffraction patterns were measured in the angular range of $2\Theta = 10–80°$ with a scanning speed of 0.02°/s. The measured diffraction patterns were analyzed using the full profile phase analysis method, which was based on the use of the POWDER CELL 2.5 software package. Phase identification was carried out using the International Center for Diffraction Data (ICDD) PDF-4 powder database.

Figure 1 shows the diffraction pattern for synthesized powder of Co-Zn ferrite. It shows a set of reflections of the spinel phase corresponding to the ferrite $Co_{0.5}Zn_{0.5}Fe_2O_4$ composition (JCPDS No. 17-8598) with a lattice parameter of 8.4206 Å.

Thermomagnetometric experiments were carried out using a Netzsch STA 449C Jupiter thermal analyzer. The temperature range of sample heating was 45–400 °C. The synthesized samples were placed in corundum (Al_2O_3) crucibles with a volume of 85 μL and were heated in an air and nitrogen atmosphere. The heating rate was 10, 20, 50 K/min. Permanent magnets were attached to the measuring cell of the analyzer to determine the magnetic phases in the samples under study. Figure 2 shows the location of the magnets during thermomagnetometric analysis. The results of the thermal analysis in the form of thermomagnetometric curves were processed using the Netzsch Proteus Analysis software.

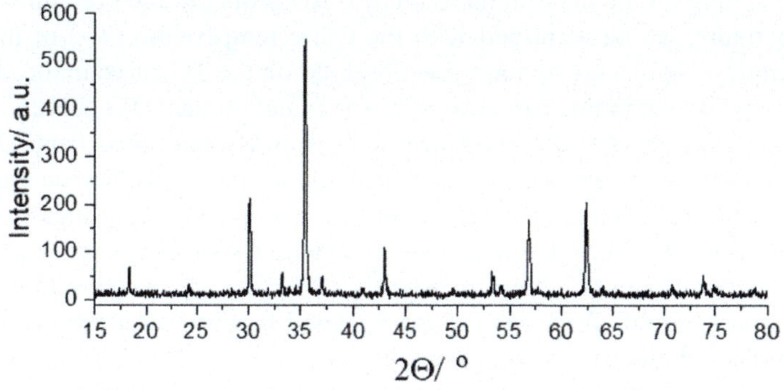

Fig. 1 XRD pattern for synthesized $Co_{0.5}Zn_{0.5}Fe_2O_4$ ferrite

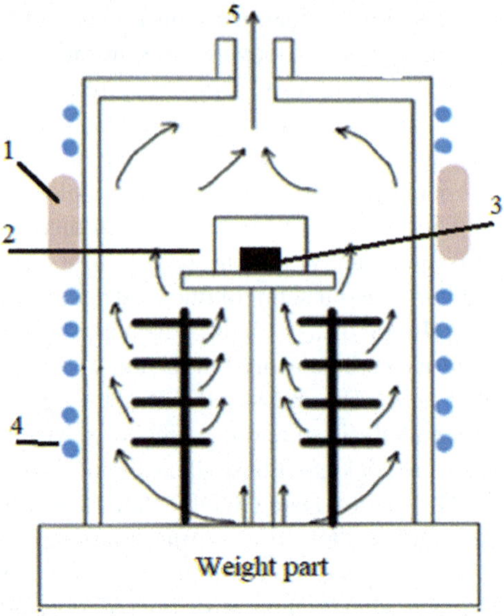

Fig. 2 Scheme of thermomagnetometric measurements, where 1 are magnets, 2 are crucibles, 3 are samples, 4 is heater

3 Experimental Part

Figures 3 and 4 show the results of the thermomagnetometric measurements, which were obtained at different regimes for $Co_{0.5}Zn_{0.5}Fe_2O_4$ ferrite. The TG and DTG curves for all the samples have a similar behavior.

The observed weight change on the TG curve in the temperature range of 160–200 °C is associated with a magnetic phase transition, that is, the transition of a ferrite sample from a magnetic to a paramagnetic state. Moreover, this transition corresponds to an intense sharp peak on the DTG curve, the temperature of which varies depending on the measurement modes. According to the literature data [24], this temperature can be identified with the Curie temperature (T_C) of the studied ferrite samples. No change in mass was observed on the TG curve in the absence of a magnetic field, and in this case, no peak was found on the DTG curve.

As was shown in [25, 26], the Curie temperature determined from DTG peak depends on the type of analyzers furnace, heating rate, duration of the magneto-phase transition, and other factors. Therefore, in thermomagnetometric measurements, a more corrective value of the Curie temperature (with the smallest experimental error) of ferrite materials can be obtained from the TG curve by crossing the tangent of the rising mass "jump" front and the extrapolated baseline after the magnetophase transition (Figs. 3 and 4).

The temperature values associated with the magnetic phase transition and obtained using the above approaches are shown in Table 1 (including the data from the cooling stage). It can be seen that the temperature of the DTG peak (T_{DTG}) depends on both the heating rate and the cooling rate during thermomagnetometric measurements.

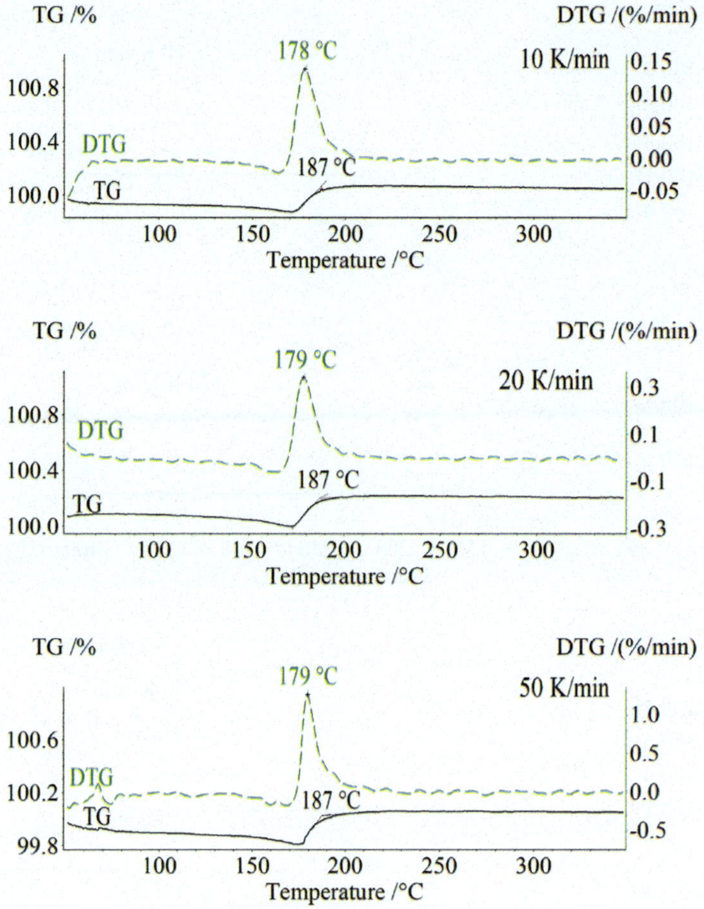

Fig. 3 Thermomagnetometric analysis obtained for $Co_{0.5}Zn_{0.5}Fe_2O_4$ at heating rate of 10, 20, 50 K/min in air

However, this parameter weakly depends on the medium composition within the experimental error.

The temperature of the DTG peak measured at the cooling stage varies in the range of 171–174 °C and it does not depend from atmosphere composition. However, the value obtained is significantly lower than the temperature measured during the heating stage. Moreover, the temperature increases from 178 to 182 °C with the increasing heating rate in nitrogen.

Also, the results showed that the temperature determined from the TG curve at the intersection of the tangent of the mass change rising front and the extrapolated baseline after the magnetic transition (T_C in Table 1) weakly depends on the heating or cooling rate, as well as atmosphere composition. However, temperature are characterized by a lower value on the cooling stage (Table 1).

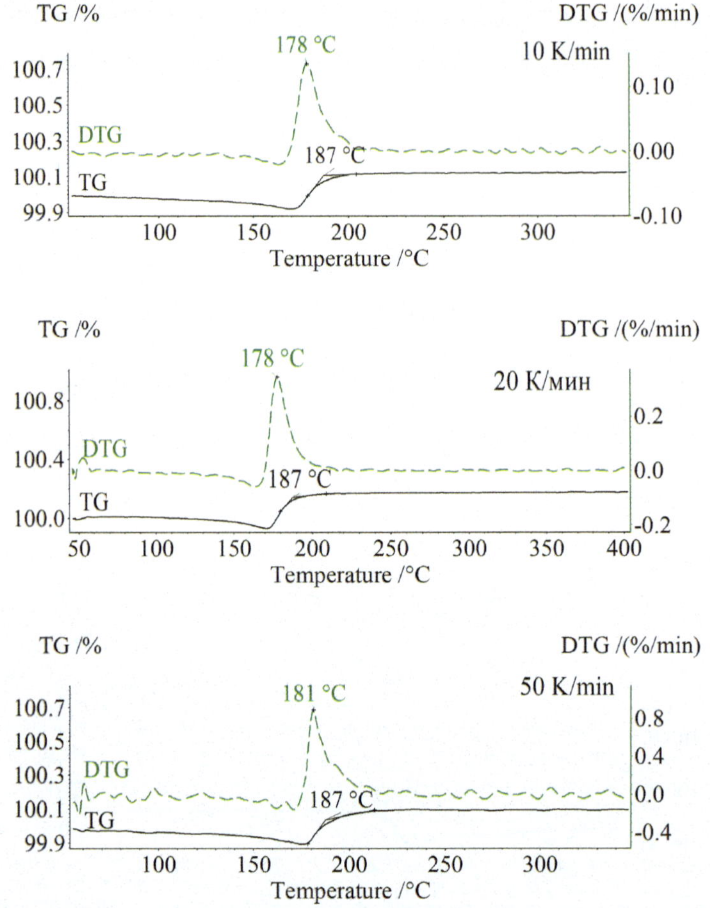

Fig. 4 Thermomagnetometric analysis obtained for $Co_{0.5}Zn_{0.5}Fe_2O_4$ at heating rate of 10, 20, 50 K/min in nitrogen

In this work, to determine the correct Curie temperature, the literature data of the Curie point for $Co_{0.5}Zn_{0.5}Fe_2O_4$ ferrite composition were studied. The data are shown in Table 2.

The analysis showed that the Curie temperature strongly depends on the particle size. The value of Curie temperature for nanopowders obtained by chemical methods is almost two times lower than that for micronpowders obtained by ceramic technology. In this case, the difference in the Curie temperature for ceramic samples is quite large and amounts to ~30 °C.

It is known that the Curie temperature also strongly depends on the chemical composition of the ferrite [31]. Any deviation from stoichiometry during the synthesis and sintering of samples can lead to a change in the Curie temperature [32, 33]. This is the reason for the large scatter observed in the Curie temperature from different

Table 1 The data from thermomagnetometric analysis for $Co_{0.5}Zn_{0.5}Fe_2O_4$ ferrite

Medium	Regime	Rate (K/min)	Δm (%)	$T_{DTG} \pm 1$ (°C)	$T_C \pm 1$ (°C) (from TG)
Air	Heating	10	0.19	178	187
		20	0.21	178	187
		50	0.24	179	187
Air	Cooling	10	0.25	173	181
		20	–	171	184
Nitrogen	Heating	10	0.19	178	187
		20	0.22	178	187
		50	0.20	181	187
Nitrogen	Cooling	10	0.23	174	181
		20	–	172	181

Table 2 Results of the Curie temperature measurements for $Co_{0.5}Zn_{0.5}Fe_2O_4$ ferrite obtained by different methods

Method of ferrite synthesis	Particle size (nm)	T_C (°C)	Refs.
Co-precipitation	9	123	[27]
	10	167	[10]
	9–11	125	[28]
	22	137	[29]
Ceramic	–	251	[14]
	–	220	[30]

literature sources. Moreover, the different T_C measuring methods with their experimental errors also play an important role in research. Comparing the results, one can notice that the Curie temperature obtained in this work for $Co_{0.5}Zn_{0.5}Fe_2O_4$ ferrite has lower values than T_C measured in [14, 30] for such ferrite obtained by ceramic technology. This may be due to the fact that the studied ferrite was obtained from ultradisperse powders prepared by mechanical grinding in a ball mill.

Thus, in thermomagnetometric measurements, the Curie temperature of ferrite can be taken from the TG curve as the temperature point of the tangent of the rising mass change front and the extrapolated base line characterizing the complete transition of the ferrite to the paramagnetic state. In this case, the heating stage can be used.

4 Conclusion

In this work, the influence of thermomagnetometric analysis conditions (heating/cooling rate, atmospheric composition) on the recorded Curie temperature of $Co_{0.5}Zn_{0.5}Fe_2O_4$ ferrite was studied.

It was found that the Curie temperature of ferrite can be determined using various approaches. These include the determination of this parameter from the derivative thermogravimetric curve, as well as from the point of intersection of the tangent of the rising mass change front and the extrapolated base line characterizing the complete transition of the ferrite to the paramagnetic state. In the first case, the Curie temperature depends on the heating and cooling rates and on the measurement mode (heating or cooling) as well as medium. In the second case, the recorded temperature depends only on the measurement modes. Therefore, this value can be used as the Curie point of the Co–Zn ferrite.

Acknowledgements This research was supported by the Russian Science Foundation (Grant no. 19-72-10078-P).

References

1. Liu, H., Ji, P., Han, X.: Rheological phase synthesis of nanosized α-$LiFeO_2$ with higher crystallinity degree for cathode material of lithium-ion batteries. Mater. Chem. Phys. **183**, 152–157 (2016)
2. Kurian, M., Thankachan, S.: Structural diversity and applications of spinel ferrites core–Shell nanostructures. A review, Open. Ceram. **8,** 100179 (2021)
3. Rani, R., Mujasam Batoo, K., Sharma, P., Anand, G., Kumar, G., Bhardwaj, S., Singh, M.: Structural, morphological and temperature dependent electrical traits of $Co_{0.9}Zn_{0.1}In_xFe_{2-x}O_a$ spinel nano-ferrites. Ceram. Int. **47**, 30902–30910 (2021)
4. Gauns Dessai, P.P., Verenkar, V.M.S.: Synthesis and characterization of $Ni_{0.7-x}Mn_xZn_{0.3}Fe_2(C_4H_2O_4)_3{\cdot}6N_2H_4$ (x = 0.1 – 0.6): a precursor for the synthesis of nickel-manganese-zinc ferrites. J. Therm. Anal. Calorim. **142**, 1399–1411 (2020)
5. Guo, J., Zhang, H., He, Z., Li, S., Li, Z.: Electrical properties and temperature sensitivity of Mo-modified $MnFe_2O_4$ ceramics for application of NTC thermistors. J. Mater. Sci: Mater. Electron. **29**, 2491–2499 (2018)
6. Inoue, A., Kong, F.: Soft magnetic materials, reference module in materials science and materials engineering (Elsevier) (2020). https://doi.org/10.1016/B978-0-12-803581-8.11725-4
7. Gao, Y., Wang, Z.: Microwave absorption and electromagnetic interference shielding properties of Li-Zn ferrite-carbon nanotubes composite. J. Magn. Magn. Mater. **528**, 167808 (2021). https://doi.org/10.1016/j.jmmm.2021.167808
8. Hajalilou, A., Kianvash, A., Lavvafi, H., Shameli, K.: Nanostructured soft magnetic materials synthesized via mechanical alloying: a review. J. Mater. Sci.: Mater. Electron. **29**, 1690–1717 (2018). https://doi.org/10.1007/s10854-017-8082-0
9. Sutradhar, S., Bandyopadhyay, A.: Modulation of magnetic and dielectric response of mullite coated Cu—substituted Co–Zn-ferrite multiphase nanocomposites. Mater. Sci. Eng: B **266**, 115079 (2021)

10. Gómez-Polo, C., Recarte, V., Cervera, L., Beato-López, J.J., López-García, J., Rodríguez-Velamazán, J.A., Ugarte, M.D., Mendonça, E.C., Duque, J.G.S.: Tailoring the structural and magnetic properties of Co-Zn nanosized ferrites for hyperthermia applications. J Magn. Magn. Mater. **465**, 211–219 (2018)
11. Andhare, D.D., Patade, S.R., Kounsalye, J.S., Jadhav, K.M.: Effect of Zn doping on structural, magnetic and optical properties of cobalt nanoparticles synthesized via. Co-precipitation method. Phys. B: Condens. Matter. **583**, 412051 (2020)
12. Muntean, C., Bozdog, M., Duma, S., Stefanescu, M.: Study on the formation of $Co_{1-x}Zn_xFe_2O_4$ system using two low-temperature synthesis methods. J. Therm. Anal. Calorim. **123**, 117–126 (2016)
13. Mondal, R., Dey, S., Majumder, S., Poddar, A., Dasgupta, P., Kumar, S.: Study on magnetic and hyperfine properties of mechanically milled $Ni_{0.4}Zn_{0.6}Fe_2O_4$ nanoparticles. J. Magn. Magn. Mater. **448**, 135–145 (2018)
14. Azouaoui, A., El haoua, M., Salmi, S., Benzakour, N., Hourmatallah, A., Bouslykhane, K.: Structural and magnetic properties of Co–Zn ferrites: density functional theory calculations and high-temperature series expansions. Comput. Condens. Matter. **23**, e00454 (2020)
15. Lanfredi, S., Genova, D.H.M., Brito, I., Lima, A., Nobre, M.: Structural characterization and Curie temperature determination of a sodium strontium niobate ferroelectric nanostructured powder. J. Solid State Chem. **184**, 990–1000 (2011). https://doi.org/10.1016/j.jssc.2011.03.001
16. Xie, Y., Fan, J., Xu, L., Zhang, X., Xu, R., Zhu, Y., Tang, R., Wang, C., Ma, C., Pi, L., Zhang, Y., Yang, H.: Unambiguous determining the Curie point in perovskite manganite with second-order phase transition by scaling method. Phys. Lett. A. **383**, 125843 (2019). https://doi.org/10.1016/j.physleta.2019.125843
17. Chen, D., Harward, I., Baptist, J., Goldman, S., Celinski, Z.: Curie temperature and magnetic properties of aluminum doped barium ferrite particles prepared by ball mill method. J. Magn. Magn. Mater. **395**, 350–353 (2015). https://doi.org/10.1016/j.jmmm.2015.07.076
18. Lysenko, E.N., Nikolaev, E.V., Surzhikov, A.P., Nikolaeva, S.A.: Kinetic analysis of lithium-titanium ferrite formation from mechanically milled reagents. J. Therm. Anal. Calorim. **239**, 122055 (2020). https://doi.org/10.1016/j.matchemphys.2019.122055
19. Lysenko, E.N., Nikolaev, E.V., Vlasov, V.A., Surzhikov, A.P.: Microstructure and reactivity of Fe_2O_3–Li_2CO_3–ZnO ferrite system ball-milled in a planetary mill. Thermochim. Acta **664**, 100–107 (2018). https://doi.org/10.1016/j.tca.2018.04.015
20. Lysenko, E.N., Surzhikov, A.P., Vlasov, V.A., Malyshev, A.V., Nikolaev, E.V.: Thermal analysis study of solid-phase synthesis of zinc- and titanium-substituted lithium ferrites from mechanically activated reagents. J. Therm. Anal. Calorim. **122**, 1347–1353 (2015). https://doi.org/10.1007/s10973-015-4849-9
21. Fischer, H.: Calibration of micro-thermal analysis for the detection of glass transition temperatures and melting points. J. Therm. Anal. Calorim. **92**, 625–630 (2008)
22. Weddle, B.J., Robbins, S.A., Gallagher, P.K.: Further studies on the use of simultaneous TM/DTA to establish magnetic transition temperatures. Pure Appl. Chem. **67**, 1843–1847 (1995)
23. Luciani, G., Costantini, A., Branda, F., Scardi, P., Lanotte, L.: Thermal evolution of ferromagnetic metallic glasses. A study using TG(M) technique. J. Therm. Anal. Calorim. **72**, 105–111 (2003)
24. Astafyev, A.L., Lysenko, E.N., Surzhikov, A.P., Nikolaev, E.V., Vlasov, V.A.: Thermomagnetometric analysis of nickel-zinc ferrites. J. Therm. Anal. Calorim. **142**, 1775–1781 (2020)
25. Gallagher, P.K.: Thermomagnetometry. J. Therm. Anal. Calorim. **49**, 33–44 (1997)
26. Lin, D.M., Wang, H.S., Lin, M.L., Lin, M.H., Wu, Y.C.: TG(M) and DTG(M) techniques and some of their applications on material study. J. Therm. Anal. Calorim. **58**, 347–353 (1999)
27. Arulmurugan, R., Vaidyanathan G., Sendhilnathan S., Jeyadevan, B.: Preparation and properties of temperature-sensitive magnetic fluid having $Co_{0.5}Zn_{0.5}Fe_2O_4$ and $Mn_{0.5}Zn_{0.5}Fe_2O_4$ nanoparticles. Phys. B: Condens. Mater. **368**, 223–230 (2005)
28. Arulmurugan, R., Vaidyanathan G., Sendhilnathan S., Jeyadevan, B.: Thermomagnetic properties of $Co_{1-x}Zn_xFe_2O_4$ (x = 0.1–0.5) nanoparticles. J. Magn. Magn. Mater. **303**, 131–137 (2006)

29. Veverka, M., Jirák, Z., Kaman, K., Maryško, M., Pollert, E., Lančok, A., Dlouhá, M., Vrastislav, S.: Distribution of cations in nanosize and bulk Co–Zn ferrites. Nanotechnology **22**, 345701 (2011). https://doi.org/10.1088/0957-4484/22/34/345701
30. Ramana Reddy, A.V., Ranga Mohan, G., Ravinder, D., Boyanov, B.S.: High-frequency dielectric behaviour of polycrystalline zinc substituted cobalt ferrites. J. Mater. Sci. **34**, 3169–3176 (1999)
31. Lee, S.W., Ryu, Y.G., Yang, K.J.: Magnetic properties of Zn^{2+} substituted ultrafine Co-ferrite grown by a sol-gel method. J. Appl. Phy. **91**, 7610 (2002). https://doi.org/10.1063/1.1452215
32. Lysenko, E.N., Astafyev, A.L., Vlasov, V.A., Surzhikov, A.P.: Analysis of phase composition of LiZn and LiTi ferrites by XRD and thermomagnetometric analysis. J. of Magn. Magn. Mater. **465**, 457–461 (2018)
33. Teo, M.L.S., Kong, L.B., Li, Z.W., Lin, G.Q., Gan, Y.B.: Development of magneto-dielectric materials based on Li-ferrite ceramics I. Densification behavior and microstructure development. Alloys Compd **459**, 557–566 (2008)

On the Ignition of Forest Areas as a Result of Man-Made and Natural Disasters

Valeriy Perminov

Abstract Publications are known where the problems of technogenic and space safety of the Earth are discussed, due to the possibility of man-made accident and collision with celestial bodies. It has been established that, as a rule, a major man-made or space disaster is accompanied by the occurrence of massive forest fires. In connection with the assessment of the environmental and climatic consequences of strong fires, it is of interest to predict the impact of this process on the state of the surface layer of the atmosphere. Due to the fact that full-scale research in such problems is simply impossible, methods of mathematical modeling are relevant. Below we consider the problem of the initial stage of the impact of a high-altitude source of radiant energy on the underlying surface of the Earth, covered with forest vegetation. The purpose of this study is to determine the size of the ignition zone and study the physical and chemical processes occurring in this case. In addition, on the basis of the results of the impact on the forests of the explosion of a celestial body, called the Tunguska meteorite, known from the scientific literature, its power is estimated.

1 Introduction

Forests cover 1254 million hectares in Russia. This represents 22.5% of the forest area worldwide. Our country is the largest producer of commercial timber. In addition, forests are a source of various types of raw materials for industry. From an ecological point of view, they play an important role in purifying the atmospheric air and enriching it with oxygen. As a result of forest fires, about 1 million hectares of forest perish annually in the Russian Federation [1]. So, in July and early August 2010, forest fires in European Russia and the Urals covered a huge area. According to the operational data of the Federal Forestry Agency, the total area covered by fire from the beginning of the year to August 3, inclusive, significantly exceeded one

V. Perminov (✉)
School of Non-Destructive Testing, National Research Tomsk Polytechnic University, 30 Lenin Ave., Tomsk 634050, Russia
e-mail: perminov@tpu.ru

E. Lysenko et al. (eds.), *Emerging Trends in Materials Research and Manufacturing Processes*, Engineering Materials, https://doi.org/10.1007/978-3-031-38964-1_2

million hectares. According to official data from the Ministry of Emergency Situations on August 4, 50 people died in forest fires. At least 130 settlements burned completely or partially. A large military base in the Moscow region burned down, the damage from the fire was approximately equal to the annual financing of the entire forestry of the country. The fire entered the territory of the Federal Nuclear Center in Sarov, Nizhny Novgorod Region, and was extinguished with great difficulty. Many large cities and entire regions of European Russia existed for weeks in life-threatening smoke, in places visibility was only a few tens of meters. This caused a partial cancellation of air traffic and hampered road traffic. According to the US National Aerospace Agency (NASA), a cloud of smoke from forest fires in the European part of Russia as of August 4, 2010 reached a width of three thousand kilometers. Smoke from forest fires in this area penetrated the stratosphere to a height of about twelve kilometers—at such an altitude it can be transported over very long distances. The fire disaster in the forests of the European-Ural region of Russia in 2010 has two main reasons. The first is an extreme drought that has engulfed most of the territory of European Russia and the Urals. The second is the lack of state forest protection. Droughts of approximately the same magnitude as 2010 occur in European Russia two or three times a century. During the twentieth century, similar droughts occurred in 1936 and 1972, both times accompanied by intense forest and peat fires. In many regions of European Russia, forests and peatlands have dried up so much that even in damp types of forests, the slightest spark is enough to start smoldering of the forest floor, quickly turning into a forest fire. The sources of almost all fires are people—cigarette butts thrown by them, bonfires left unattended, etc. For the future, the main thing that needs to be done to prevent the catastrophic development of forest fires is to create a full-fledged state forest guard, the only responsibility of which would be to protect forests [1, 2]. In addition, it is necessary to recreate a unified system of aviation protection of forests and extinguishing large forest fires (an analogue of the former Federal State Institution Avialesookhrana), including to ensure the possibility of a quick transfer of qualified forces and equipment from low-burning regions to heavily burning ones. In 2010, the Ministry of Emergency Situations of the Russian Federation feared that forest fires in areas affected by the Chernobyl accident could cause radiation contamination. According to Minister of Emergency Situations Shoigu, "in the event of a fire there, radionuclides may rise along with combustion products and a new zone may appear where there is such pollution," the minister noted.

Crown fires are the most dangerous type of fires. They account for up to 70% of the burnt area. It should be noted that the mechanisms and conditions for the occurrence of various types of forest fires have not yet been fully elucidated. Putting out forest fires requires a lot of effort and money, and in the vast majority of cases it is ineffective or impossible [1, 2].

In addition, large forest fires may occur as a result of man-made and natural disasters [1–6]. These are explosions resulting in the formation of large fireballs, including explosions of rocket propellants (including solid and liquid propellants), explosions of chemical products, rupture of vessels with a subsequent explosion of a vapor cloud in an open volume, burning of liquids in open tanks, detonation of high explosives substances and nuclear explosions—all these phenomena can lead to the formation of fireballs. A fireball is a cloud of burning gas or vapor that rises above the earth's surface and is characterized by large thermal radiation at considerable distances. The danger of fireballs is due to their high temperature, large size and the ability to move as a result of the action of the wind. Therefore, they can serve as sources of ignition and ignite various combustible materials along the trajectory of their movement. The occurrence of fires in large areas, including forest areas, can lead to such a phenomenon as a "fire storm" [2, 4–6].

This phenomenon can also occur as a result of natural disasters (for example, the explosion of the Tunguska meteorite) [3]. The formation of a fireball is accompanied by a powerful shock wave. The fireballs are hot enough to cause radiant damage. In this case, combustible materials are ignited and a burn effect is exerted on a person. Increased attention to this problem is also due to the impact of large combustion centers on the surface layer of the atmosphere, which is accompanied by climatic (a decrease in the temperature of the environment due to smoke in the territories causes the death or later ripening of agricultural crops) and environmental consequences [2].

Due to the fact that experimental methods for studying forest fires are expensive and do not allow for a complete physical modeling of this phenomenon, theoretical research methods are of interest [7–10]. So, as studies have shown, the method of mathematical modeling makes it possible to adequately describe the state of forest biogeocenosis and the surface layer of the atmosphere during forest fires. For example, on the basis of numerical analysis with the help of a computer, it is possible to study the process of transition from a ground fire to a crown fire in the absence and presence of wind and ignition of the forest canopy by light radiation.

A literature review of publications on mathematical modeling of forest fires shows that they have not yet been studied in full [5, 6, 9, 11]. Therefore, it is of interest to study the process of transition from a ground fire to a crown fire, since it can become the basis for the development of new ways to combat forest fires. The study of the spread of forest fires, especially in forests subject to radioactive contamination, will also contribute to the development of new methods for the prevention and control of this phenomenon. Consideration of issues related to the problem of ignition of forests under the influence of radiation will help in the study of this phenomenon in order to develop safety measures in case of possible man-made and natural disasters.

The purpose of the paper is to formulate and theoretically study the problems of the occurrence of large-scale forest fires under the influence of light radiation on forest tracts resulting from natural and man-made disasters, considering the two-temperature environment.

The research technique was based on the numerical solution of one-dimensional, two-dimensional and three-dimensional unsteady Reynolds equations for describing turbulent flow, taking into account diffusion equations for chemical components and energy conservation equations for gaseous and condensed phases. In the numerical solution, splitting by physical processes was used, that is, first the hydrodynamic pattern of the flow and distribution of the desired scalar functions was calculated, and then the equations of chemical kinetics for the volume fractions of the phases were solved and the chemical sources for the scalar functions were taken into account. To obtain discrete analogues, the control volume method was used.

2 Physical and Mathematical Model

Let the source of radiant energy be at a height H from the Earth's surface (Fig. 1). Since its dimensions are small compared to the radius of the Earth, we will consider it a point source of radiation, D is the distance from the center of the source to the current point of the forest surface, h is the height of the forest, 0 is the epicenter, r* is the radius of the ignition zone. An intense radiant flux $q_R(r, t)$ acts on the upper boundary $z = h$ of the forest area, which weakens with distance from the epicenter 0. The maximum intensity of the source is reached at t = to, then it decays to zero according to the data on $q_R(r,t)$ [4–6], which can be approximated as follows

$$q_R(r,t) = \frac{t_p P_m \sin L}{4\pi D^2}\begin{cases} t/t_m, & t < t_m \\ \exp(-k_0(t/t_m - 1)), & t \geq t_m \end{cases},$$

$$t_0 = 0.032 w_0^{0.5},\ P_m = 16.8 \times 10^{12} w_0^{0.5}\ \text{J/s}. \quad (1)$$

where t_m is the time of maximum heat release of the radiation source, s; D is the distance from the center of the radiation source to the forest canopy, m; t_p is the atmospheric transmittance; P_m is the maximum value of the light pulse at time t_m, J/s; L is the angle between the direction of the radiation flux density vector and the upper boundary of the vegetation cover; w_0 is source power, k_0 is approximation factor ($k_0 = 0.75$).

The entry of radiant energy into the vegetation cover ($z_o \leq z \leq h$) causes heating of forest combustible materials, evaporation of moisture and subsequent thermal decomposition of solid material with the release of volatile pyrolysis products. Then the latter burn in the atmosphere interacting with atmospheric oxygen. Due to the presence of gravity, the heated volumes of air begin to float up, so the processes of volumetric ignition of forest vegetation are, in general, associated with the hydrodynamics of the flow. Due to the fact that on the periphery of the epicenter of the

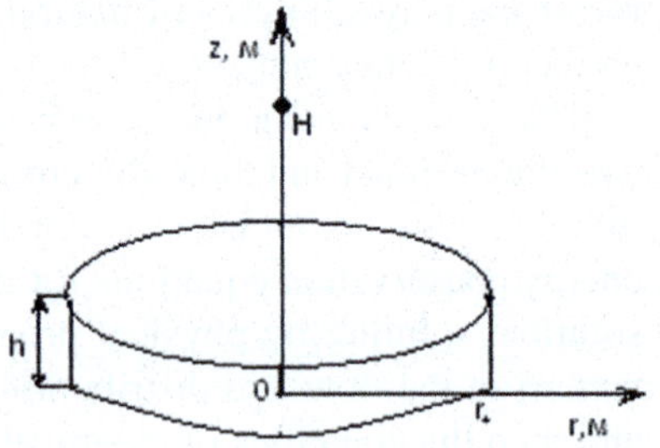

Fig. 1 Scheme of the process of ignition of the forest

explosion the intensity of the radiant flux into the forest canopy is low, no ignition occurs there. Thus, during the action of the radiation source, a zone of initial ignition of the forest mass with radius r_* is formed. Ideally, it has the shape of a circle in plan. Its subsequent development is determined by the interaction of ascending flows with the wind field, since they carry solid burning elements into the surface layer of the atmosphere and scatter over the surrounding area, as well as meteorological and geographical conditions in a given area.

For the purposes of this study, we will assume that the wind speed in the atmosphere is relatively low and energy is mainly transferred due to radiation. This allows us to consider the problem in an axisymmetric formulation. Since a combination of various physical factors accompanies the process of forest ignition, it is expedient to describe it at various levels of complexity. The hierarchy of physical models, including more complex ones, makes it possible to evaluate the role of individual factors that are omitted in order to simplify the description of the phenomenon.

Physical and mathematical models of heat and mass transfer during forest fires were considered in detail in previous papers, for example, in [7–12]. Here, the main physical assumptions and ideas about the object of study necessary for understanding the mathematical model will be indicated. It is believed that: (1) the flow is symmetrical about the vertical axis z, which has its origin in the center of the region under consideration (Fig. 1) and is directed vertically upwards, (2) the flow is of a developed turbulent nature and molecular transport is neglected compared to turbulent, (3) the density of the gas phase does not depend on pressure due to the smallness of the flow velocity compared to the speed of sound [13, 14], (4) the forest canopy is considered to be a non-deformable medium. We assume that the forest canopy can be modeled by a homogeneous two-temperature multiphase porous reacting medium [1, 9, 12]. The temperature of the condensed (solid) T_s and gas T phases is distinguished. The first includes dry organic matter, moisture, condensed pyrolysis products and the mineral part of forest combustible materials. In the gas phase, we will isolate only the components necessary to describe the combustion reaction, that is, the mass concentrations c_α ($\alpha = 1$—oxygen, 2—combustible pyrolysis products, 3—other inert components, including water vapor). The solid phase, which is a combustible material (needles and thin twigs up to 6 mm), water in a liquid-drop state and condensed pyrolysis products, does not have its own velocity and its volume fraction, compared to the gas phase, can be neglected in the corresponding equations, since per unit volume of the forest is <0.5 kg [7, 8]. From the point of view of hydrodynamics, this porous medium, however, resists with force any movement of air masses [1, 7–9, 15]. It is believed that the medium is in local thermodynamic equilibrium [16]. Turbulent convective transport due to the action of gravity is described using the Reynolds equations [9, 17].

We will also take into account the physicochemical processes occurring in the forest canopy, the rates of chemical reactions of which R_i ($i = 1, 2, 3, 5$) as a function of temperature and other parameters are considered known. Radiation is the determining mechanism of energy transfer in this case. In the forest canopy, it is mainly the solid phase that absorbs, reflects, and re-radiates. To describe transport in such a specific continuous medium, we use the diffusion approximation [17, 18].

This is justified, since the mean free path of radiation in the canopy is $l_R << l_o$, $l_0 \sim$ 10–15 m, $l_R \sim 1$ m (l_o is the characteristic height scale).

Let's direct the z-axis vertically upwards, the r-axis along the surface of the earth (Fig. 1). To describe the processes of heat and mass transfer in a volume with forest vegetation, general conservation laws for a multiphase medium are used [1, 7–9, 19]. Since any movement of air flows in the atmosphere is turbulent, the Reynolds equations are used to describe them. The formulated problem in a cylindrical coordinate system ($n = 1$) is reduced to solving the following equations [11, 20]

$$\frac{\partial \rho}{\partial t} + \frac{1}{r^n}\frac{\partial}{\partial r}(r^n \rho v) + \frac{\partial}{\partial z}(\rho w) = \dot{m}; \tag{2}$$

$$\frac{\partial}{\partial t}(\rho v) + \frac{1}{r^n}\frac{\partial}{\partial r}(r^n \rho v^2) + \frac{\partial}{\partial z}(\rho v w) = -\frac{\partial p}{\partial r} + \frac{1}{r^n}\frac{\partial}{\partial r}(-r^n \rho \overline{v'^2}) + \frac{\partial}{\partial z}(-\rho \overline{v'w'}) - \rho s c_d v\sqrt{v^2 + w^2}; \tag{3}$$

$$\frac{\partial}{\partial t}(\rho w) + \frac{1}{r^n}\frac{\partial}{\partial r}(r^n \rho v w) + \frac{\partial}{\partial z}(\rho w^2) = -\frac{\partial p}{\partial z} + \frac{1}{r^n}\frac{\partial}{\partial r}(-r^n \rho \overline{v'w'}) + \frac{\partial}{\partial z}(-\rho \overline{w'^2}) - \rho s c_d w\sqrt{v^2 + w^2} - \rho g; \tag{4}$$

$$\frac{\partial}{\partial t}(\rho c_p T) + \frac{1}{r^n}\frac{\partial}{\partial r}(r^n \rho v c_p T) + \frac{\partial}{\partial z}(\rho w c_p T) = \frac{1}{r^n}\frac{\partial}{\partial r}(-r^n \rho c_p \overline{v'T'}) + \frac{\partial}{\partial z}(-\rho c_p \overline{w'T'}) + k_g(cU_R - 4\sigma T^4) + q_5 R_5 + \alpha_v(T_s - T); \tag{5}$$

$$\frac{\partial}{\partial t}(\rho c_\alpha) + \frac{1}{r^n}\frac{\partial}{\partial r}(r^n \rho v c_\alpha) + \frac{\partial}{\partial z}(\rho w c_\alpha) = \frac{1}{r^n}\frac{\partial}{\partial r}(-r^n \rho \overline{v'c'_\alpha}) + \frac{\partial}{\partial z}(-\rho \overline{w'c'_\alpha}) - R_{5\alpha}, \alpha = 1, 2; \tag{6}$$

$$\frac{1}{r^n}\frac{\partial}{\partial r}\left(\frac{r^n c}{3k}\frac{\partial U_R}{\partial r}\right) + \frac{\partial}{\partial z}\left(\frac{c}{3k}\frac{\partial U_R}{\partial z}\right) - kcU_R + 4\sigma\left(k_g T^4 + k_s T_s^4\right) = 0; \quad k = k_g + k_S; \tag{7}$$

$$\sum_{i=1}^{4} \rho_i c_{pi} \varphi_i \frac{\partial T_s}{\partial t} = q_3 R_3 - q_2 R_2 + k_S\left(cU_R - 4\sigma T_s^4\right) + \alpha_V (T - T_s); \tag{8}$$

$$\rho_1 \frac{\partial \varphi_1}{\partial t} = -R_1, \rho_2 \frac{\partial \varphi_2}{\partial t} = -R_2, \rho_3 \frac{\partial \varphi_3}{\partial t} = \alpha_c R_1 - \frac{M_c}{M_1} R_3, \rho_4 \frac{\partial \varphi_4}{\partial t} = 0; \tag{9}$$

$$\sum_{\alpha=1}^{3} c_\alpha = 1, \; p_e = \rho RT \sum_{\alpha=1}^{3} \frac{c_\alpha}{M_\alpha},$$

$$\dot{m} = (1 - \alpha_c)R_1 + R_2 + \frac{M_c}{M_1}R_3,\ R_{51} = -R_3 - \frac{M_1}{2M_2}R_5,\ R_{52} = \nu(1 - \alpha_c)R_1 - R_5. \quad (10)$$

$$R_1 = k_1\rho_1\varphi_1 \exp\left(-\frac{E_1}{RT_s}\right),\ R_2 = k_2\rho_2\varphi_2 T_s^{-0.5} \exp\left(-\frac{E_2}{RT_s}\right),$$

$$R_3 = k_3\rho\varphi_3 s_\sigma c_1 \exp\left(-\frac{E_3}{RT_s}\right),\ R_5 = k_5 M_2\left(\frac{c_1 M}{M_1}\right)^{0.25}\frac{c_2 M}{M_2}T^{-2.25}\exp\left(-\frac{E_5}{RT}\right).$$

The initial and boundary conditions for the system of Eqs. (2) and (10) have the following form:

$$t = 0 : v = 0, w = 0, T = T_e, c_\alpha = c_{\alpha e}, T_s = T_e, \varphi_i = \varphi_{i e}; \quad (11)$$

$$r = 0 : v = 0, \frac{\partial w}{\partial r} = 0, \frac{\partial T}{\partial r} = 0, \frac{\partial c_\alpha}{\partial r} = 0, \frac{\partial U_R}{\partial r} = 0; \quad (12)$$

$$r = r_e : \frac{\partial v}{\partial r} = 0, \quad \frac{\partial w}{\partial r} = 0, \frac{\partial c_\alpha}{\partial r} = 0, \frac{\partial T}{\partial r} = 0, \quad \frac{c}{3k}\frac{\partial U_R}{\partial r} + \frac{c}{2}U_R = 0; \quad (13)$$

$$z = z_0 : \frac{\partial v}{\partial z} = 0, \frac{\partial w}{\partial z} = 0, \frac{\partial T}{\partial z} = 0, \frac{\partial c_\alpha}{\partial z} = 0, : -\frac{c}{3k}\frac{\partial U_R}{\partial z} + \frac{c}{2}U_R = 0; \quad (14)$$

$$z = z_e : \frac{\partial v}{\partial z} = 0, \frac{\partial w}{\partial z} = 0, \frac{\partial T}{\partial z} = 0, \frac{\partial c_\alpha}{\partial z} = 0, \quad \frac{c}{3k}\frac{\partial U_R}{\partial z} + \frac{c}{2}U_R = 2q_R(r, z). \quad (15)$$

where z_o is the level of roughness; r and z are coordinates, z is measured from the ground cover, and r from the axis of symmetry; $R_1 - R_5$, $R_{5\alpha}$ are mass rates of pyrolysis of forest combustible materials, evaporation of moisture, combustion of condensed and volatile pyrolysis products, formation of soot and ash and formation of α—components of the gas-dispersed phase; c_{pi}, ρ_i, φ_i are specific heat capacities, densities and volume fractions of the i-th phase (1—dry organic matter. 2—water in liquid—droplet state, 3—condensed pyrolysis products, 4—mineral part, 5—gas phase); T, T_s are temperature of the gas and condensed phases; c_α are mass concentrations ($\alpha = 1$—oxygen, 2—CO, 3—inert air components); p is the pressure; U_R is the radiation energy density; σ is the Stefan-Boltzmann constant; k is the radiation attenuation factor; k_g, k_S are absorption coefficients for gas-dispersed and condensed phases; α_V is the phase exchange coefficient, q_i, E_i, k_i are thermal effects, activation energies and pre-exponents of pyrolysis reactions, evaporation, combustion of coke and volatile pyrolysis products; s_σ is the specific surface of the element of forest combustible materials; M_α, M_c, M—are the molecular weights of individual components of the gas phase, carbon and air mixture; s, c_d are the specific surface of phytomass and empirical forest canopy coefficient; c is the speed of light; v, w are the velocity projections on the r and z axes; $\alpha_{c\ \nu,}$ are the coke number and mass fraction of combustible gases in the mass of volatile pyrolysis products; $\dot{m}$—mass rate of formation of the gas-dispersed phase; g is the free fall acceleration. Indexes "o" and "*e*" refer to the values of the functions in the combustion source and at a large distance from the fire zone, respectively. The superscript prime "′" refers to the ripple component of the given quantity. The system of Eqs. (2)–(10) describes the transport processes in the forest area, which includes the space between the underlying surface and the lower boundary of the forest canopy, the forest canopy, and the space above the canopy. Thermodynamic, thermophysical and structural characteristics correspond to the forest combustible materials of a pine forest [7–9] and are numerically equal to $E_1/R = 9400$ K, $k_1 = 3.36 \times 10^4$ s^{-1}, $q_1 = 0$, $E_2/R = 6000$ K, $k_2 = 6 \times 10^5$ s^{-1}, $q_2 = 3 \times 10^6$ J/kg, $E_3/R = 10^4$ K, $k_3 = 10^3$ s^{-1}, $q_3 = 1.2 \times 10^7$ J/kg, $E_5/R = 11{,}500$ K, $k_5 = 3 \times 10^{13}$, $q_5 = 10^7$ J/kg, $c_{p1} = 2000$, $c_{p2} = 4180$, $c_{p3} = 900$, $c_{p4} = 1000$, $c_p = 1000$ J/(kg K), $s_\sigma = 1000$ m^{-1}, $SCd = 0.1$, $\alpha_c = 0.06$, $\rho_4\varphi_4 = 0.08$ kg/m^3, $\nu = 0.7$, $\rho_1 = 500$, $\rho_2 = 1000$, $\rho_3 = 200$, $\rho_e = 1.2$ kg/m^3, $c_{2e} = 0$, $\varphi_{3e} = 0$, $p_e = 10^5$ n/m^2, $T_e = 300$ K, $c_{1e} = 0.23$.

A locally-equilibrium turbulence model was used [1, 11]. The system of Eqs. (2)–(10) contains terms related to turbulent convection and needs to be closed. The components of the turbulent stress tensor $\rho\overline{v_i'v_j'}$, as well as turbulent heat and mass fluxes $\overline{\rho v_j' c_p T'}$, $\overline{\rho v_j' c_\alpha'}$, are written in terms of mean flow gradients according to [1, 11].

$$-\rho\overline{v_i'v_j'} = \mu_t\left(\frac{\partial v_i}{\partial x_j} + \frac{\partial v_j}{\partial x_i}\right) - \frac{2}{3}K\delta_{ij},$$
$$-\overline{\rho v_j' c_p T'} = \lambda_t\frac{\partial T}{\partial x_j},\ -\overline{\rho v_j' c_\alpha'} = D_t\frac{\partial c_\alpha}{\partial x_j},$$
$$\lambda_t = \mu_t c_p/\mathrm{Pr}_t,\quad \rho D_t = \mu_t/Sc_t,\ \mu_t = c_\mu \rho K^2/\varepsilon$$

where K is the kinetic energy of turbulence, v_i and v_i' are the components of the average velocity and the pulsation component of the velocity [19] in projection onto the x_i axis; μ_t, λ_t, D_t are the coefficients of turbulent dynamic viscosity, turbulent thermal conductivity and diffusion; Pr_t. Sc_t are the turbulent Prandtl and Schmidt numbers; δ_{ij} are the Kronecker symbols. Taking into account the notation adopted in (2)–(10) $x_1 = r$, $x_2 = z$, $v_1 = v$, $v_2 = w$. According to [1, 15], where ε is the rate of dissipation of turbulent kinetic energy, c_μ is a constant. Due to the fact that the determination of the coefficient of turbulent dynamic viscosity is associated with difficulties due, for example, to arbitrariness in the choice of initial and boundary conditions for the equation of kinetic energy of turbulence, an approximate method of closure based on the Prandtl mixing path hypothesis is of interest, which actually means an equilibrium approximation (balance of generation and dissipation) for the turbulence kinetic energy equation. The details of this approach for the two-dimensional plane case are presented in [1, 17]. For the axisymmetric case under consideration, we assume that in the equation for the kinetic energy of turbulence in a cylindrical coordinate system, one can neglect the nonstationary and convective terms, as well as the terms of the diffusion of turbulent kinetic energy. Then, from the right side of the equation, we can obtain an expression for the kinetic energy of turbulence and, according to [15], the coefficient of turbulent dynamic viscosity.

$$\mu_t = \rho\left(\frac{C_\mu}{C_1}\right)^{\frac{3}{2}} l^2 \Big\{ 2\Big[\left(\frac{\partial w}{\partial z}\right)^2 + \left(\frac{\partial v}{\partial r}\right)^2 + \left(\frac{v}{r}\right)^2\Big] + \left(\frac{\partial w}{\partial r} + \frac{\partial v}{\partial z}\right)^2$$
$$- \frac{2}{3}\Big[\frac{\partial w}{\partial z} + \frac{1}{r}\frac{\partial}{\partial r}(rv)\Big]^2 - \frac{g}{T Pr_T}\frac{\partial \theta}{\partial z}\Big\}^{\frac{1}{2}},$$

Following [1, 15], it was believed that

$$l = zk_t/(1 + 2.5z\sqrt{c_d s/h}),$$

where $k_t = 0.4$ is Karman's constant, C_1, C_μ are constants, h is the size of the forest canopy. Equation of μ_t corresponds to the local-equilibrium model of turbulence [1, 15].

In the forest canopy the process of attenuation and emission of radiation is influenced by the elements of forestry materials (needles, branches, etc.), as well as dispersed particles of soot and ash formed during combustion. Since typical forestry elements are cylinders with a diameter d_s and a length l_S uniformly oriented in all directions, according to [8, 9], the absorption and scattering coefficients in the forest canopy can be determined by the equations:

$$k_S = \frac{\alpha_S}{d_S}\varphi_S, \quad \beta_S = \frac{\rho_S}{d_S}\varphi_S,$$

where φ_S—volume fraction of forest materials; α_S, ρ_S—absorption and reflection coefficients for one element of fuel. For the dispersed phase, it is assumed that soot and ash particles have the shape of a sphere with diameters d_c and d_3. According to [8, 9], the absorption and scattering coefficients have the following form

$$k_g = \frac{\alpha_c}{d_c}\varphi_c + \frac{\alpha_3}{d_3}\varphi_3, \quad \beta_g = \frac{\rho_c}{d_c}\varphi_c + \frac{\rho_3}{d_3}\varphi_3,$$

where α_c, α_3, and ρ_c, ρ_3 are the absorption and reflection coefficients for individual soot and ash particles, respectively; φ_c, φ_3 are the volume fractions of soot and ash.

The phase heat transfer coefficient α_V was chosen on the basis of data on heat transfer between the element of forest combustible material (needle, thin twig) and the environment. It was believed that a typical FCM element is a cylinder with radius r_s and length l, which has different angles of inclination to the vertical. The convective heat flux from the gas phase to the solid is defined as

$$q = \alpha(T_S - T),$$

where α is the heat transfer coefficient, T_S, T are the temperatures of the cylinder and the gas phase, respectively. Then, as a result of convective heat transfer per unit volume of the gas-dispersed phase, the heat influx is determined by the equation

$$Q_k = \alpha n F_c (T_S - T) = A_S (T_S - T),$$

where n is the number of particles (inclusions) per unit volume of a multiphase medium, which, in accordance with the fundamental assumptions, does not change during a fire by determining the volume fraction of the condensed phase, we have

$$n = \varphi_S / V_S, \varphi, \quad \varphi_S = \sum_{i=1}^{4} \varphi_i, \; V_S = \pi r_S^2 l.$$

where φ_i is the volume fraction of the condensed phase, and V_s is the volume of inclusions. The coefficient α_V is equal to $\alpha n F_c$, F_c is the surface area of the cylinder, equal to $2\pi rSl$. To determine A_S, we substitute the value n

$$\alpha_V = \alpha_s S, \text{ где } S = 2\varphi_S / r_S,$$

where S is the specific surface of the inclusions. Thus, to determine α_V, it is necessary to know the heat transfer coefficient between the gas phase and the FCM element. In this case, it should be taken into account that when heated, moisture evaporates from the FCM and decomposition into pyrolysis products occurs, which can then ignite. Knowing the numerical value of the Nusselt criterion, we find the heat transfer coefficient for a single needle $\alpha = Nu/d$. Here d is the diameter of the FCM element, λ is the coefficient of its thermal conductivity. Using the results of [1, 11], an equation for the Nusselt number was get in [1]

$$Nu = Nu_{00} \varepsilon_0 \, \varepsilon_{Tu} \, \varepsilon_\delta \, \varepsilon_n \, \varepsilon_B \, \varepsilon_g,$$

where Nu_{oo} is the Nusselt number without taking into account the mass velocity of blowing from the surface of the fuel and fuel element and the degree of turbulence. ε_0 is a multiplier allowing to take into account the effect of injection of gaseous decomposition products of fuel and fuel elements, ε_{Tu} is the factor for taking into account flow turbulence, ε_δ is the factor for taking into account the angle of attack δ (the angle between the velocity vector and the axis of the cylinder), flow, ε_B is the factor for taking into account the influence of vibrations of a typical FCM element, ε_g is the factor for taking into account the effect of combustion reactions in the vicinity of the FCM element. As noted above, when element of FCM is heated, moisture evaporates and a gas-dispersed mixture is released during pyrolysis. As a result, the value of the interfacial heat transfer coefficient decreases. According to [1], the injection correction was taken into account by the following equation

$$\alpha_V = \alpha S - \gamma c_p \dot{m},$$

where γ is a parameter that characterizes the ratio of the molecular weights of the gas in the environment and blown out during heat exchange [1].

3 Results of Numerical Calculations

The system of Eqs. (2)–(10) with initial and boundary conditions (11)–(15) for numerical integration was reduced to a discrete form using the Patankar-Spalding control volume method [21]. The grid equations arising in the course of discretization were solved using the ADI or SIP [22] methods. The velocity and pressure fields were matched iteratively using the SIMPLE algorithm [21, 23]. As a result of numerical integration, the fields of temperatures, mass concentrations of the gas phase components, volume fractions of the solid phase components, as well as vector velocity fields at different times were obtained. Figures 2 and 3 shows the distributions of the temperature fields of the gas and solid phases at the time when the ignition zone has its maximum dimensions. These figures describe the processes taking place in the forest canopy. The explosion energy in this calculation was 8.4×10^{13} J, and its center was at a height H = 260 m above the Earth's surface. The greatest heating of the medium is observed in the central part of the region under consideration.

Here, masses of heated gases rise up due to the action of the Archimedean force. The vector flow pattern is shown in Fig. 4. It should be noted that the wind is directed from the periphery to the center of the ignition region. However, the wind speed at the time of ignition is not high and does not exceed 0.5 m/s for the considered times. The distribution of mass concentrations of combustible volatile products of pyrolysis is inextricably linked with the ignition process, the most characteristic of which is carbon monoxide (Fig. 5) and oxygen (Fig. 6), which is consumed in the combustion process. The zone of formation of volatile combustible products of pyrolysis moves into the depths of the forest canopy as it warms up. In the central part of the region under consideration, CO burns out and, as a result, a decrease in

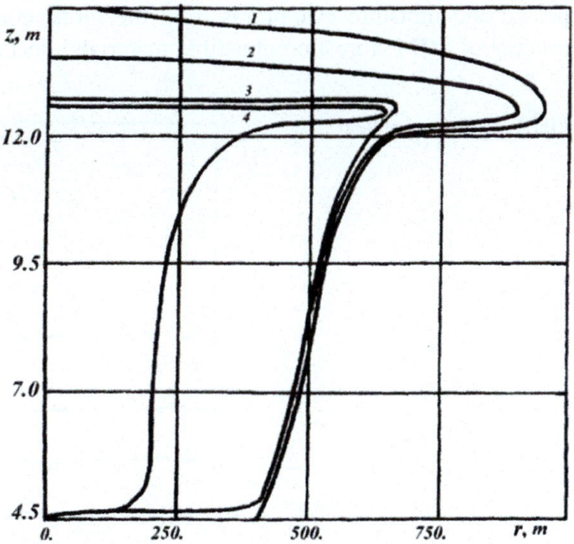

Fig. 2 Gas phase isotherms: $1 - \overline{T} = 2.6, 2 - \overline{T} = 3.5, 3 - \overline{T} = 7, 4 - \overline{T} = 10; \overline{T} = T/Te$

Fig. 3 The condensed phase isotherms $1 - \overline{T}_S = 10$, ⇄ $2 - \overline{T}_S = 4$, $3 - \overline{T} = 3$

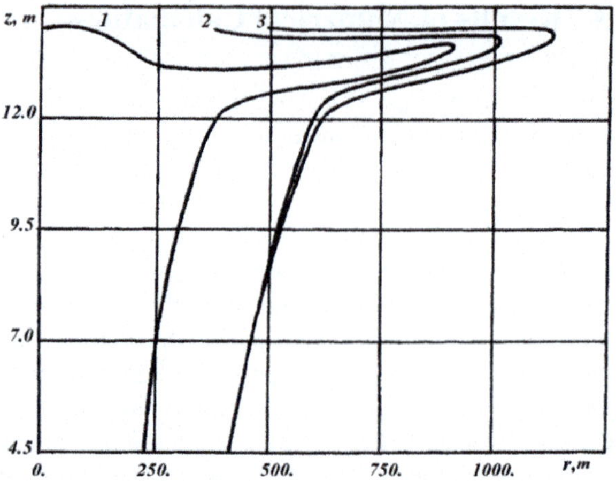

the amount of oxygen occurs. A curious pattern is observed at the boundary of the ignition region (see Fig. 5), where there is non-monotonicity in the distribution of volatile combustible pyrolysis products (CO). This is directly related to the ignition process, as a result of which they burn out in the area where ignition takes place.

At the same time, outside the ignition zone, due to the fact that the power of the light pulse depends on the distance to the center of the explosion and decreases with time according to Eq. (1) [4–6], the amount of energy supplied is insufficient for ignition to occur, and there remain burnt gaseous and condensed products of pyrolysis. As a result of the action of the Archimedean force, the heated gaseous pyrolysis products float above the forest canopy. It should be noted that the ignition process has a gas-phase multi-stage character. First, the solid and gas phases are heated and moisture evaporates from the forest combustible materials. Then the process of FCM (forest combustible materials) decomposition into condensed and

Fig. 4 Vector speed field

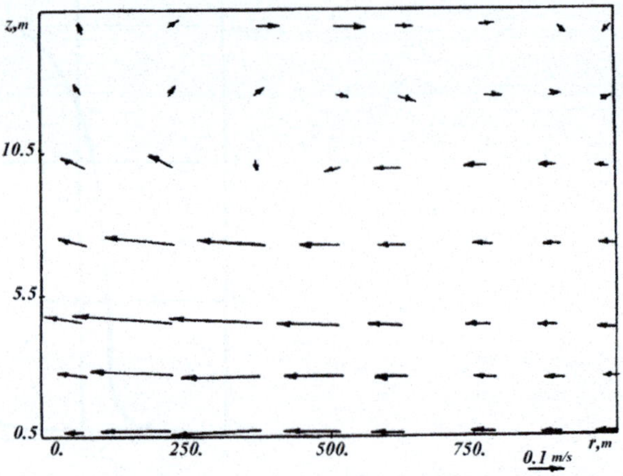

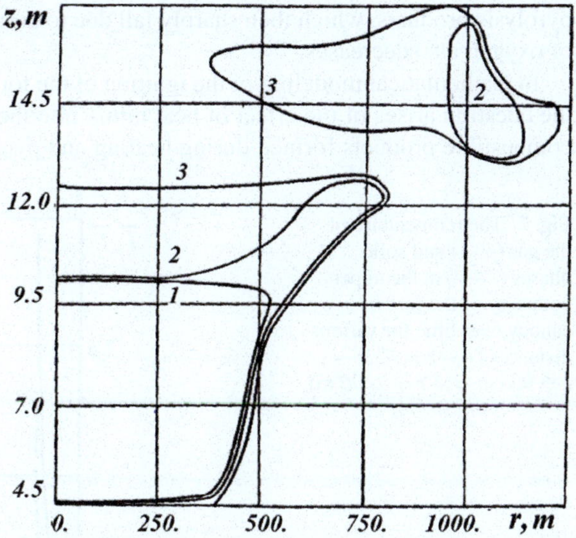

Fig. 5 Distribution of lines of equal mass concentration of volatile combustible pyrolysis products: $1 - \overline{c}_2 = 1$, $2 - \overline{c}_2 = 0.01$, $3 - \overline{c}_2 = 0.001$; $\overline{c}_\alpha = c_\alpha / c_{1e}$, $c_{1e} = 0.23$

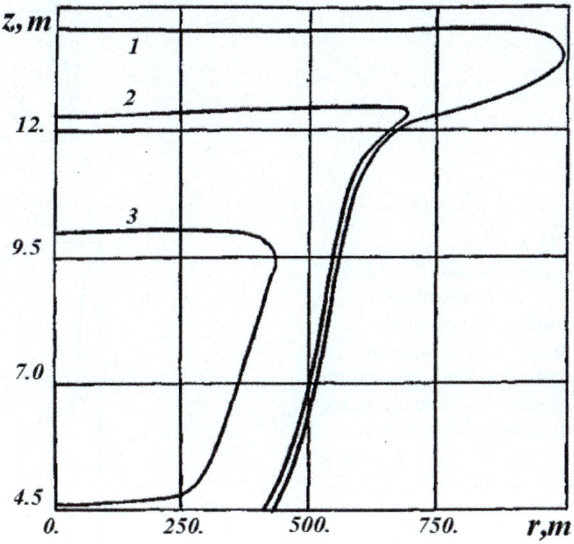

Fig. 6 Distribution of lines of equal mass concentration of oxygen: $1 - \overline{c}_1 = 0.7$, $2 - \overline{c}_1 = 0.06$, $3 - \overline{c}_1 = 0.001$

volatile pyrolysis products begins, the latter of which ignite at the upper boundary of the forest canopy.

Figure 8 shows the distributions of mass concentrations of oxygen (solid lines) and volatile combustible pyrolysis products (dashed lines) for the situation presented in Fig. 7. The modes of combustion and ignition (dashed lines 1–2 in Fig. 8) are characterized by the formation of maxima in the release of gaseous combustible pyrolysis products, which then sharply fall due to burnout. Accordingly, the amount of oxygen also decreases.

In mathematical modeling of the ignition of the forest canopy by a radiation flux, the question arises of the effect of heat influx into the gas phase due to the gaseous combustible products formed during heating and pyrolysis. In this case, the energy

Fig. 7 The temperature of the gas (-----) and solid phases (——) at the upper boundary of the forest canopy over time for various distances (*1* – ***r*** = *0*, *2* – ***r*** = *900* m, *3* – ***r*** = *1200* m)

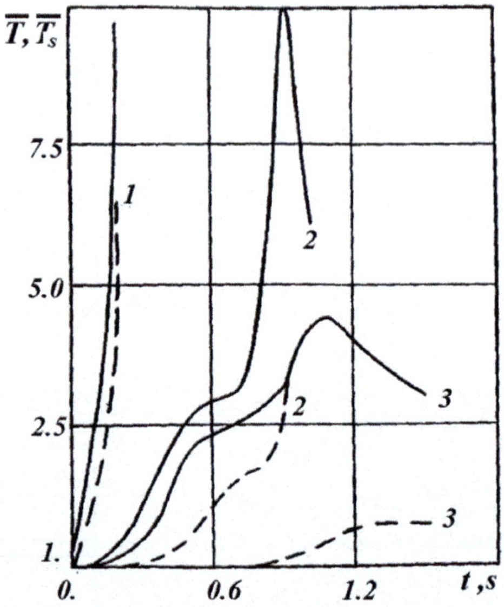

Fig. 8 The mass concentrations of oxygen (———) and volatile combustible pyrolysis products (--------) at the upper boundary of the forest canopy over time for various distances (*1* – *r* = *0*, *2* – *r* = *900* m, *3* – *r* = *1200* m)

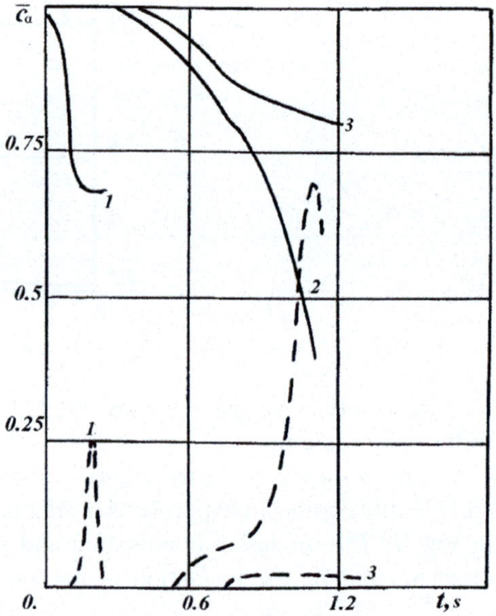

equation for the gas phase will have the form (5). Evaluation of the numerical value of the value of the additional term $c_p(T_s - T)\,\dot{m}$ on the right side of the energy equation for the gas phase showed that its value during the entire ignition process is two orders of magnitude lower than the value of the term $\alpha_V(T_s - T)$, which describes the heat transfer of the phases. During ignition, heat release from the combustion reaction of combustible pyrolysis products also exceeds the value of the considered term by a factor of 10^2. The maximum dimensions of the ignition zone have not changed. Thus, this source term can be neglected in the calculations.

4 Physical and Mathematical Model of Ignition in a Quasi-One-Dimensional Formulation

In the previous section. The statement of the problem of the ignition of forests under the influence of light radiation as a result of a man-made disaster (using the example of a nuclear explosion) in a two-dimensional axisymmetric approximation is presented. The radiant heat flux from high-altitude source q_r, falling on the forest canopy, significantly exceeds the energy transfer in the horizontal direction. In this regard, this problem can be considered in a quasi-one-dimensional approximation, taking into account the change in the main functions along the vertical coordinate z for different distances r from the explosion epicenter. As the estimates show, until the moment of ignition, the following inequalities are satisfied:

$$q_{kx1}, << q_r, q_{Rx1}, << q_r, q_{kx2}, << q_r, q_{Rx2}, << q_r,$$

where q_{kx2},, q_{kx2}, q_{Rx1}, q_{Rx2} are the projection densities of convective q_k and radiative q_R heat fluxes on the horizontal x_1 and x_2 axes in the forest phytocenosis. Therefore, we will assume that the effect of wind speed on the ignition of fuel and fuel can be neglected, because energy is mainly transferred by radiation.

In contrast to a collisional catastrophe, when considering the process of an air explosion, one can not consider the thermal effect of radiation from a charge at the flight stage and take into account the radiant flux after the explosion as from a stationary source. It follows from the above estimates that this problem can be considered in a quasi-one-dimensional formulation for both man-made (for example, a nuclear explosion) and space catastrophe.

For simplicity, we assume, as indicated above, that the horizontal component of the wind speed in the forest canopy has little effect on the ignition process. This assumption is correct, since the radiant heat flux is much greater than the convective one. For this reason, we consider the process to be quasi-one-dimensional, that is, we assume that all parameters depend on the time t and the vertical coordinate z.

Taking into account the assumptions made and the conservation laws for a multiphase multicomponent reacting medium, mathematically, the problem is reduced to solving the following equations [27–30].

$$\frac{\partial \rho}{\partial t} + \frac{\partial}{\partial z}(\rho w) = \dot{m}; \tag{16}$$

$$\frac{\partial}{\partial t}(\rho w) + \frac{\partial}{\partial z}(\rho w^2) = -\frac{\partial p}{\partial z} + \frac{\partial}{\partial z}(-\rho \overline{w'^2}) - \rho s c_d w^2 - \rho g; \tag{17}$$

$$\frac{\partial}{\partial t}(\rho c_p T) + \frac{\partial}{\partial z}(\rho w c_p T) = \frac{\partial}{\partial z}(-\rho c_p \overline{w'T'}) + k_g(cU_R - 4\sigma T^4) + q_5 R_5 + \alpha_v(T_s - T); \tag{18}$$

$$\frac{\partial}{\partial t}(\rho c_\alpha) + \frac{\partial}{\partial z}(\rho w c_\alpha) = \frac{\partial}{\partial z}(-\rho \overline{w'c'_\alpha}) - R_{5\alpha}, \alpha = 1, 2; \tag{19}$$

$$\frac{\partial}{\partial z}\left(\frac{c}{3k}\frac{\partial U_R}{\partial z}\right) - kcU_R + 4k_g \sigma T^4 + 4k_S \sigma T_s^4 = 0; \tag{20}$$

$$\sum_{i=1}^{4} \rho_i c_{pi} \varphi_i \frac{\partial T_s}{\partial t} = q_3 R_3 - q_2 R_2 + k_S(cU_R - 4\sigma T_s^4) + \alpha_V(T - T_s); \tag{21}$$

$$\rho_1 \frac{\partial \varphi_1}{\partial t} = -R_1, \rho_2 \frac{\partial \varphi_2}{\partial t} = -R_2, \rho_3 \frac{\partial \varphi_3}{\partial t} = \alpha_c R_1 - \frac{M_c}{M_1} R_3, \rho_4 \frac{\partial \varphi_4}{\partial t} = 0; \tag{22}$$

The initial and boundary conditions for the system of Eqs. (16)–(22) have the following form

$$t = 0 : w = 0, T = T_e, c_\alpha = c_{\alpha e}, T_s = T_e, \varphi_i = \varphi_{i e}; \tag{23}$$

$$z = z_0 : \frac{\partial w}{\partial z} = 0, \frac{\partial T}{\partial z} = 0, \frac{\partial c_\alpha}{\partial z} = 0, -\frac{c}{3k}\frac{\partial U_R}{\partial z} = \frac{\varepsilon}{2(2-\varepsilon)}(4\sigma T_s^4 - cU_R); \tag{24}$$

$$z = h : \frac{\partial v}{\partial z} = 0, \frac{\partial w}{\partial z} = 0, \frac{\partial T}{\partial z} = 0, \frac{\partial c_\alpha}{\partial z} = 0, \frac{c}{3k}\frac{\partial U_R}{\partial z} + \frac{c}{2}U_R = 2q_R(r, z). \tag{25}$$

All designations and the way of closing the terms associated with turbulent characteristics are given above in [8, 9].

5 Results of Numerical Calculations

At the first stage, within the framework of the quasi-one-dimensional approximation in the mathematical formulation (16)–(25), calculations were carried out to determine the maximum sizes of the ignition zones of forest areas as a result of man-made disasters (using nuclear explosions as an example). The results of these calculations were compared with the data obtained in the framework of a two-dimensional axisymmetric formulation.

Figure 9 shows the maximum dimensions of the ignition zones for various powers of nuclear explosions obtained from two-dimensional and quasi-one-dimensional formulations. Comparison of the results obtained by the mathematical statement in the two-dimensional approximation (2)–(15) and one-dimensional (16)–(25) showed that a satisfactory agreement is achieved between the maximum sizes of the ignition zones of forest tracts. However, it should be noted that the ignition radii obtained in the framework of the quasi-one-dimensional approximation are approximately 5–10% larger than those calculated using the two-dimensional formulation. This increase in the maximum ignition radii is explained by the fact that in the one-dimensional case the energy transfer in the horizontal direction is not taken into account.

This provides a more intense heating of the FCM at a given point in the forest canopy, and ultimately leads to the formation of ignition zones at greater distances from the explosion epicenter than in the two-dimensional case. The above comparison justifies the possibility of using a quasi-one-dimensional approximation to determine the maximum dimensions of the ignition zones of forest tracts under the influence of light radiation. Obviously, the use of a one-dimensional system of equations in calculations is more economical in terms of programming and computer time. As a result of numerical integration, the fields of temperatures, mass concentrations, and volume fractions of phases are obtained at different times depending on the z coordinate for different distances from the explosion epicenter. Figures 10 and 11 show the distribution of relative temperatures and mass concentrations of oxygen and combustible gas at the upper boundary of the forest canopy over time. Figure 12 shows the temperature distribution of the gas and condensed phases, as well as the mass concentrations of oxygen and volatile combustible pyrolysis products in the forest canopy, for the epicenter of the explosion. These calculations were carried out for an explosion energy of 105×10^{15} J. In contrast to a nuclear explosion, the fraction of light energy emitted according to [3] was 10%. As the analysis of these graphs shows, the following sequence of events takes place. Initially, under the action of light radiation, the solid phase is heated. Moisture evaporates and dry

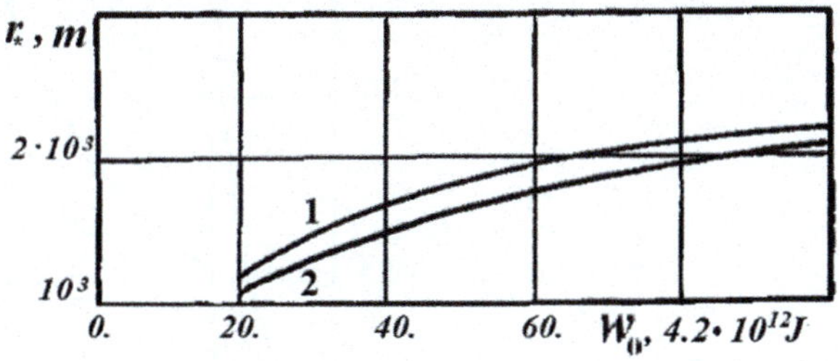

Fig. 9 The results of the calculation of the radii of the ignition zones: 1—calculation according to the quasi-one-dimensional setting, 2—calculation according to the two-dimensional axisymmetric setting; $H = 260$ m, $\rho_o = 0.2$ kg/m^3, $W = 0.6$, $T_e = 300$ K, $k = 0.3$

FCM is heated with further decomposition into coke and volatile pyrolysis products. Presented in Fig. 10 temperature distributions (solid lines) of the condensed (1) and gas (2) phases and concentrations (dashed lines) of the oxygen component (curve 1) and gaseous pyrolysis products (2) along the height of the forest canopy correspond to the moment of ignition. It can be seen from the graph that the temperature of the solid phase exceeds the gas temperature, since the heating in the forest canopy is carried out due to the external radiation flux.

At the moment of ignition at the upper boundary of the forest canopy, the temperature of the gas phase rises due to the ignition of volatile pyrolysis products, the concentration of which has decreased, and their maximum has shifted into the canopy into the unignited zone. The oxygen concentration also decreased. An analysis of the curves presented in Fig. 11 allows us to state that, as in the case of the transition from a ground fire to a crown fire, two ignition modes take place—normal and burning [11, 24–26]. The burning ignition mode is realized near the epicenter of the explosion and is characterized by elevated temperatures of the solid phase (up to 4000 K) (curve 1 in Fig. 11), a large burnout of condensed and gaseous pyrolysis products, as well as a significant consumption of oxygen. In essence, in this case forced combustion of FCM occurs under the action of a powerful radiant flux. The heat released during the oxidation of pyrolysis products does not significantly affect the temperature of the gas phase in the forest canopy. Moreover, due to heat exchange with the condensed phase, the temperature of the gas phase increases and can be higher than the adiabatic combustion temperature (Fig. 10 dashed curve I). The concentration of oxygen decreases over time, and the concentration of combustible gas first increases, and then sharply decreases due to burnout.

The maximum temperature of the gas phase and the minimum of oxygen concentration are realized at the upper boundary of the forest canopy (Figs. 10, 11 and 12).

Fig. 10 The temperatures of the gas (--------) and solid phases (——) at the upper boundary of the forest canopy over time for various distances (*1* − $r = 0$, *2* − $r = 7.5$ km, *3* − $r = 9$ km)

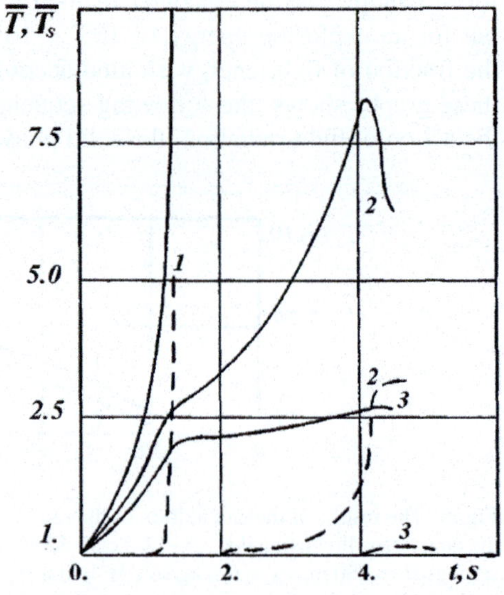

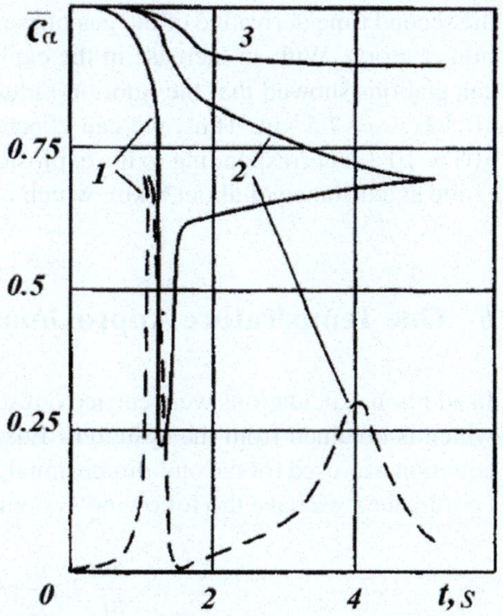

Fig. 11 The mass concentrations of oxygen (———) and volatile combustible pyrolysis products (--------) at the upper boundary of the forest canopy over time for various distances (*1*-$r = 0$, *2*-$r = 7.5$ km, *3* $- r = 9$ km)

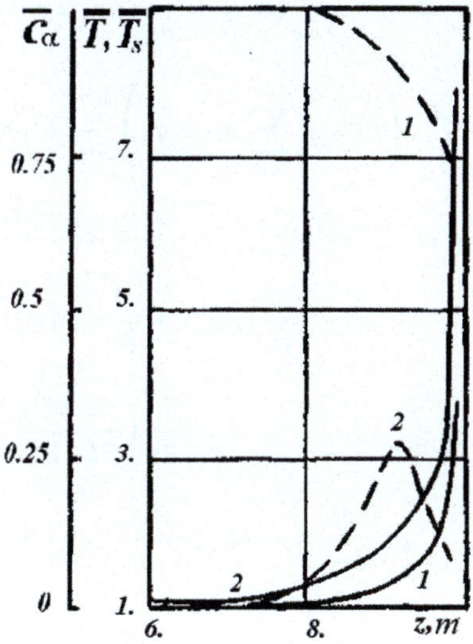

Fig. 12 Temperature distribution (——) (1) and (2) of phases and mass concentrations (--------) of oxygen (1) and volatile combustible pyrolysis products (2) in the forest canopy

As the distance from the explosion epicenter increases to $r = 7.5$ km, the temperature at the upper boundary of the forest canopy decreases (Fig. 10, curve 2) due to a decrease in the density of the radiant heat flux. Therefore, gaseous products of pyrolysis are less intensively oxidized and less oxygen is consumed (curve 2 in Fig. 11). In this case, the dependence of the temperature of the gas phase on time (see curve 2 in Fig. 10) has two inflection points. As mentioned above, in this case, the normal ignition mode is implemented. In this case, the condition of equality of the second time derivative of the gas phase temperature to zero was used as the ignition criterion. With an increase in the explosion energy to 126×10^{15} J, numerical calculations showed that the ignition radius increases to 10 km, and at wo $= 84 \times 10^{15}$ J, $r_* = 7.5$ km. Thus, we can conclude that with an energy release of about 105×10^{15} J, corresponding to the explosion of the Tunguska meteorite, tree crowns ignite at a distance of about 8 km, which is consistent with known data [31–36].

6 One Temperature Approximation

In addition, calculations were carried out according to the single-temperature model, which is obtained from the system of Eqs. (16)–(22) at $A_S \to \infty$ [37]. The energy equation was used for the one-dimensional case, that is, in projection onto the vertical coordinate z we have the following system of equations

$$\frac{\partial \rho}{\partial t} + \frac{\partial}{\partial z}(\rho w) = \dot{m}; \tag{26}$$

$$\frac{\partial}{\partial t}(\rho w) + \frac{\partial}{\partial z}(\rho w^2) = -\frac{\partial p}{\partial z} + \frac{\partial}{\partial z}(-\rho \overline{w'^2}) - \rho s c_d w^2 - \rho g; \tag{27}$$

$$\left(\rho c_p + \sum_{i=1}^{4} \rho_i c_{pi} \varphi_i\right)\frac{\partial T}{\partial t} + \frac{\partial}{\partial z}(\rho w c_p T) = \frac{\partial}{\partial z}(-\rho c_p \overline{w'T'}) + k(cU_R - 4\sigma T^4) - q_2 R_2 + q_3 R_3 + q_5 R_5; \tag{28}$$

$$\frac{\partial}{\partial t}(\rho c_\alpha) + \frac{\partial}{\partial z}(\rho w c_\alpha) = \frac{\partial}{\partial z}(-\rho \overline{w' c'_\alpha}) - R_{5\alpha}, \alpha = 1, 2; \tag{29}$$

$$\frac{\partial}{\partial z}\left(\frac{c}{3k}\frac{\partial U_R}{\partial z}\right) - k(cU_R - 4\sigma T_s^4) = 0; \tag{30}$$

$$\rho_1 \frac{\partial \varphi_1}{\partial t} = -R_1, \rho_2 \frac{\partial \varphi_2}{\partial t} = -R_2, \rho_3 \frac{\partial \varphi_3}{\partial t} = \alpha_c R_1 - \frac{M_c}{M_1} R_3, \rho_4 \frac{\partial \varphi_4}{\partial t} = 0; \tag{31}$$

$$\sum_{\alpha=1}^{3} c_\alpha = 1, p_e = \rho R T \sum_{\alpha=1}^{3} \frac{c_\alpha}{M_\alpha}, \tag{32}$$

The initial and boundary conditions for the system of Eqs. (26)–(32) have the following form

$$t = 0: w = 0, T = T_e, c_\alpha = c_{\alpha e}, \varphi_i = \varphi_{ie}; \tag{33}$$

$$z = z_0 :, \frac{\partial w}{\partial z} = 0, \frac{\partial T}{\partial z} = 0, \frac{\partial c_\alpha}{\partial z} = 0, -\frac{c}{3k}\frac{\partial U_R}{\partial z} = \frac{\varepsilon}{2(2-\varepsilon)}(4\sigma T^4 - cU_R); \tag{34}$$

$$z = h : \frac{\partial v}{\partial z} = 0, \frac{\partial w}{\partial z} = 0, \frac{\partial T}{\partial z} = 0, \frac{\partial c_\alpha}{\partial z} = 0, \frac{c}{3k}\frac{\partial U_R}{\partial z} + \frac{c}{2}U_R = 2q_R(r, z). \tag{35}$$

All designations and the method of closing the terms associated with turbulent characteristics are given above. The terms associated with turbulent diffusion of mass, momentum and energy are written in terms of mean flow gradients. The local equilibrium model of turbulence was used [38].

The system of Eqs. (26)–(32) with initial and boundary conditions (33)–(35) was numerically integrated using the Patankar-Spaulding method [21]. The calculations used an algorithm for automatic selection of the time step and a non-uniform grid in space. In the course of the numerical solution of the problem, the fields of temperatures, volume fractions of phases, and mass concentrations of components at different times were obtained. An analysis of the graphs shown in Figs. 13 and 14 shows that initially, under the action of light radiation, the forest fuel is heated with further evaporation of moisture and decomposition into coke and volatile pyrolysis products. An analysis of the curves in Fig. 13 allows us to state that there are two ignition modes, normal and burning [24–26]. The burning ignition mode, as in the previous cases, is realized near the epicenter of the explosion and is characterized by an increased temperature and significant burnout of gaseous and condensed pyrolysis products. Essentially, in this situation, the forced combustion of FCM occurs under the action of a powerful radiant heat flux. The temperature in this case may be higher than the adiabatic combustion temperature. At some distance from the epicenter of the explosion (Fig. 13, curve 2), the so-called normal ignition mode is realized [24–26, 39], which is characterized by relatively low burnout of pyrolysis products.

The equality of the second time derivative of the temperature at the upper boundary of the forest canopy was also used as an ignition criterion. Using this condition, the

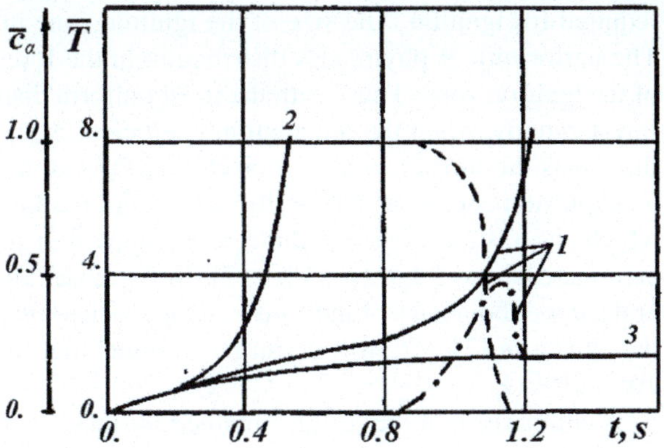

Fig. 13 Changes in temperature and mass concentrations of oxygen and volatile combustible pyrolysis products at the upper boundary of the forest canopy over time for various distances: 1 – $r = 0$, 2 – $r = 6$ km; 3 – $r = 10$ km; ——— $\overline{T}$, -------- $\overline{c}_1$, - • - • - • - • - $\overline{c}_2$; $w_0 = 105 \times 10^{15}$ J

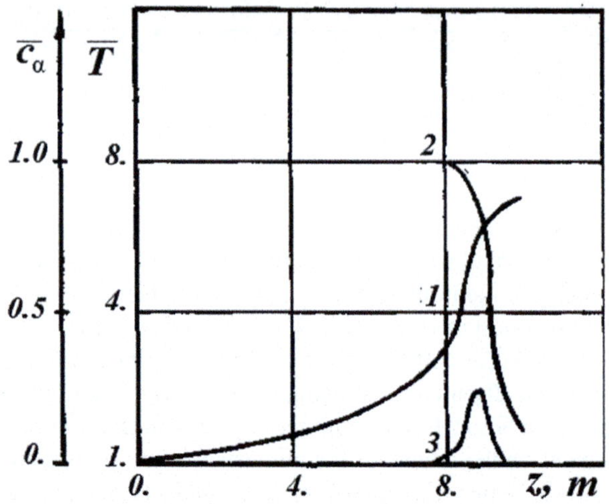

Fig. 14 Distribution of temperature $\overline{T}$ (1) and mass concentrations of oxygen $\overline{c}_1$ (2) and volatile combustible pyrolysis products $\overline{c}_2$ (3) in the forest canopy

ignition time was determined only for the normal mode. With further distance from the epicenter of the explosion, the forest canopy does not ignite.

Initially, ignition is carried out at the upper boundary of the forest canopy. Here, the maximum temperature value is realized, and the mass concentration of oxygen decreases, since it is consumed for the oxidation of gaseous pyrolysis products. The maximum concentration of the latter shifts deeper into the forest canopy (Fig. 14).

Due to the heterogeneity of the distribution of forest fuels in the canopy and their moisture content, it is of interest to study the effect of these characteristics on the size of the ignition zone under the action of light radiation (Figs. 15, 16 and 17). The explosion energy in these calculations was 84×10^{15} J (curves 1), 105×10^{15} J (curves 2) and 126×10^{15} J (curves 3). As expected, with an increase in moisture content from 0.1 to 0.66, the size of the ignition zone decreased (Fig. 15).

With an increase in the bulk of FCM in the forest canopy, the effective heat capacity of the FCM layer increases and, therefore, a more powerful light pulse is required for ignition. The size of the ignition zone in this case decreases (Fig. 16). The uniformity of phytomass distribution in the forest canopy also affects the size of the ignition zone. Thus, with the most uniform distribution of fuel and fuel in the forest canopy, when the attenuation coefficient is $k = 0.375\ m^{-1}$, the radiant heat flux heats the forest canopy relatively evenly and the radius of the ignition zone at an explosion energy of 105×10^{15} J is about 6.5 km. With an increase in k to 0.8, which ensures a decrease in the forest canopy transparency, the size of the ignition area increases to 9 km (see Fig. 17). In this case, almost all the energy is retained at the upper boundary of the forest canopy, where ignition takes place. Based on the data given in [31, 32, 40], it can be assumed that the explosion energy was about 105×10^{15} J.

A comparison of the results obtained for these two models showed that they give some differences in the size of the ignition zones. They amounted to about 12%. Thus, at the explosion energy $w_o = 105 \times 10^{15}$ J, $r_* = 7.5$ km for the one-temperature

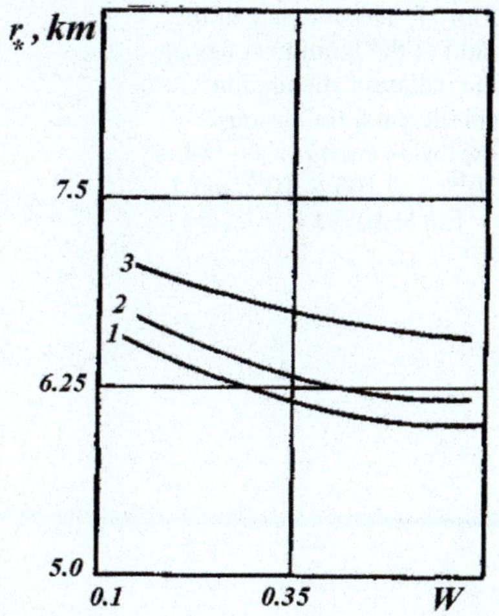

Fig. 15 Dependence of the radii of ignition zones on the moisture content of forest fuel *W* for various explosion energies

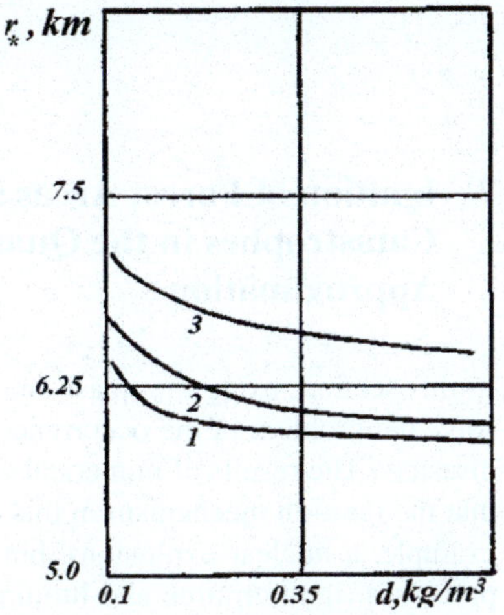

Fig. 16 Dependence of the radii of the ignition zones on the bulk of fuel and fuel *d* for various explosion energies

model and $r_* = 8.5$ km for the two-temperature model. To confirm the adequacy of the mathematical model for the phenomenon under study, it is of interest to introduce the concept of tree burn as the degree of pyrolysis of their fuels (~20% of the initial volume fraction), then numerical calculations using a two-temperature model show that this phenomenon occurs at a distance of about 16 km. That is, heating, moisture evaporation and the process of FCM pyrolysis without their ignition take place there, which is consistent with the data of papers [31–33]. Thus, the two-temperature model, in contrast to the one-temperature model, allows one to study such a phenomenon as tree burns.

Fig. 17 Dependence of the radii of the ignition zones on the radiation attenuation coefficient k for various explosion energies: 1 – 84 × 10^{15}, 2 – 105 × 10^{15} and 3 – 126 × 10^{15} J

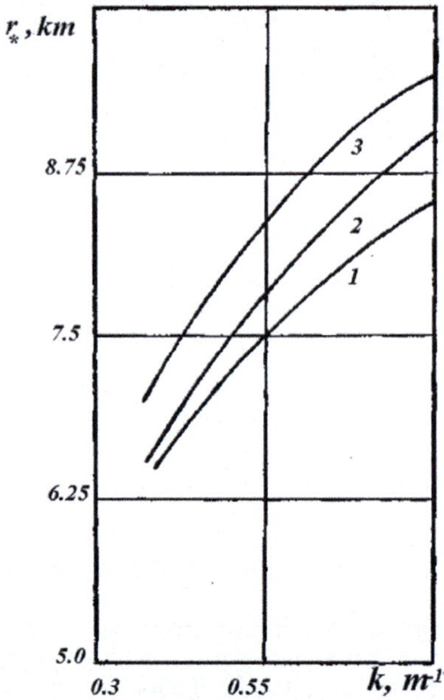

7 Ignition of Forest Areas as a Result of Collisional Catastrophes in the Quasi-One-Dimensional Approximation

In this section, using the quasi-one-dimensional approximation, we formulate and solve the problem of the occurrence of massive forest fires as a result of collisional disasters. The results of numerical calculations are presented, from which it follows that the ignition mechanism in this case is the same as for man-made disasters (for example, a nuclear explosion), but the quantitative characteristics of the ignition processes (ignition time and limiting conditions, the shape of the ignition zone on the underlying surface) differ significantly. This is due to the difference in the mechanisms of energy release in the surface layer of the atmosphere during man-made and collisional disasters. It is known that when bodies enter the dense layers of the atmosphere at a high supersonic speed, a powerful ballistic shock wave arises in the vicinity of the frontal part of the body, the gas temperature in which takes on very high values. As a result, sublimation of the material of the celestial body occurs and huge thermal stresses arise in it. Therefore, the celestial body, ultimately, either collapses in the atmosphere of the Earth or another planet, or the remains of this body fall to the surface of the planet with the formation of a crater. During the entire flight time, part of the kinetic energy of the celestial body is emitted in the atmosphere and the Earth's surface, including forest phytocenoses, is heated. As a rule, the dimensions of a celestial body are small compared to the radius of the Earth and the thickness of the atmospheric layer. Therefore, a celestial body flying in the Earth's atmosphere can be considered a point source of radiation. We will assume that the celestial body ceases to exist in the Earth's atmosphere as a result of an explosion, as is commonly believed in relation to the Tunguska celestial body [36].

Let the source of radiant energy at the initial moment of time be at a height H from the Earth's surface (Fig. 18). In this figure, R_0 is the distance from the source center to the current point of the forest area surface, h is the height of the forest area, 0 is the epicenter of the explosion, which is taken as the origin of the Cartesian coordinate system. The upper boundary $z = h$ of the forest area is affected by the density of the radiant flux $q(r,t)$, which is determined through the brightness of the celestial body. At the stage of the flight of the Tunguska meteorite, according to the data of [36], we have

$$I = 0{,}5 \ \ C_{\mathcal{H}}\rho V^3 S_m, \tag{36}$$

where I is the brightness of the Tunguska fireball, J/s; V—its speed at any point on the trajectory, m/s; C_H is the fraction of the kinetic energy of the oncoming flow, which is converted into radiation; Sm is the area of the midsection of the body. After the explosion of a celestial body, the radiant flux is determined according to [4, 27, 28]. The entry of radiant energy into the vegetation cover (0 ≤ z ≤ h causes heating of forest combustible materials, evaporation of moisture and subsequent thermal decomposition of solid material with the release of volatile pyrolysis products. Then the latter burn out in the atmosphere interacting with atmospheric oxygen. Due to the presence of gravity heated volumes of air begin to float up, therefore, the processes of heat and mass transfer during the ignition of forest vegetation, in a forested area, are generally associated with the flow of air masses.

As the above estimates prove, the effect of wind speed on the ignition of fuel and fuel in the atmosphere can be neglected, since energy is mainly transferred by radiation. In contrast to collisional catastrophes, in an air explosion of a nuclear bomb it

Fig. 18 Scheme of the fall of the Tunguska meteorite

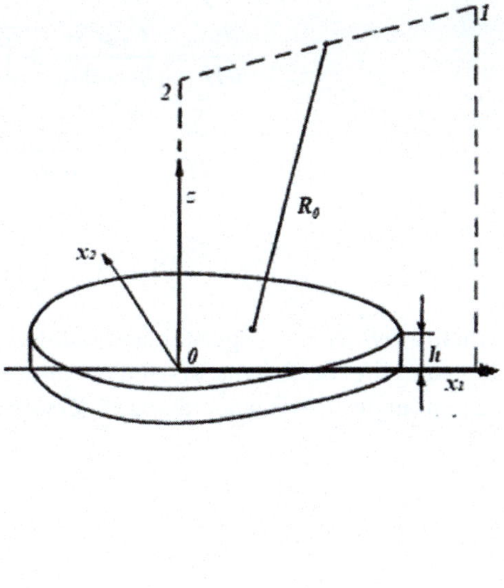

makes no sense to consider the flight stage of the radiation source and it can be considered motionless. It follows from the above estimates that this problem can be considered in a quasi-one-dimensional setting, both for space and man-made disasters. The forest canopy will be considered as a homogeneous, two-temperature, reacting, non-deformable medium [1]. The temperature of the condensed (solid) T_S and gaseous T phases are distinguished. The first is dry organic matter. moisture. condensed pyrolysis products and the mineral part of forest combustible materials. In the gas phase, we will isolate only the components c_α necessary to describe the combustion reaction ($\alpha = 1$—oxygen, 2—combustible pyrolysis products of forest combustible materials, 3—other components, including water vapor). The solid phase, which is a combustible forest material (needles and thin twigs up to 6 mm in diameter), does not have its own velocity, so it can be interpreted as a kind of porous medium [1]. The volume fraction of the dispersed phase (smoke particles) is small up to the moment of ignition and, compared with the gas phase, it can be neglected. Radiation is the determining mechanism of energy transfer in this case. In the forest canopy, it is mainly the solid phase that absorbs, reflects, and re-radiates. Let us assume, for simplicity, that the solid phase is a gray medium. To describe transport in such a specific continuous medium, we will use the diffusion approximation [41, 42]. This is justified. Since the mean free path of radiation in the canopy is $l_R << 10$ (l_0 is the characteristic height scale). $lo \sim 10$–15 m, $l_R \sim 1$ m. Due to the fact that the luminosity of the fireball is a function of its speed, it is necessary to write down a system of equations to describe the flight of a celestial body in the Earth's atmosphere. It is assumed that the motion of the cosmic body occurs quasi-stationary and is described by the system of equations given in [35, 36]:

$$\frac{dV}{dH} = \frac{C_x \rho V S_T}{2m \sin\alpha_0} - \frac{g}{V}, \tag{37}$$

$$\frac{dm}{dH} = 6\frac{C_x \rho V^2 S_T}{\sin\alpha_0}, \tag{38}$$

$$\frac{d\alpha}{dH} = \frac{C_Y \rho\, S_T}{2m \sin\alpha_0} + \left(\frac{1}{R_z} - \frac{g}{V^2}\right) ctg\alpha_0, \tag{39}$$

$$\frac{dt}{dH} = -\frac{1}{V \sin\alpha_0}, \frac{d\ell}{dH} = -\frac{1}{\sin\alpha_0}, \tag{40}$$

$$S_m = \pi R_T^2, R_T = \left(\frac{3m}{4\pi\rho_T}\right)^{\frac{1}{3}}, \sigma_0 = \frac{\Lambda}{2C_x Q},$$

where V, m, R_T, ρ_S—speed, mass, radius, and density of the meteor body, C_x, C_Y—drag and lift coefficients, H—height above the Earth, t—time, α_o—trajectory inclination angle, l—distance along trajectories, ρ is the density of the atmosphere at height H, g is the acceleration of gravity, R_z is the radius of the Earth, σ_o is the ablation coefficient, Λ is the heat transfer coefficient, Q is the specific ablation energy of the meteorite substance.

According to [36], it was assumed that the body has a spherical shape ($C_Y = 0$) and a constant drag coefficient $C_x = 1$ during hypersonic flight. The mass loss was calculated using a model that assumes the evaporation of matter and the entrainment of the entire evaporated mass. The ablation coefficient varied in the range from 0.5×10^{12} to 5×10^{12} s^2/cm^2. The starting point of the trajectory under consideration was assumed to be $H = 60$ km, the explosion point was at a height of $H = 6.5$ km. According to [36], it was assumed that $\alpha_o = \text{const} = 40°$. The initial values of the velocity and mass of the meteorite, depending on the estimated density of the meteorite matter, were chosen from the data given in [36].

To describe the processes of heat and mass transfer in a volume with forest vegetation, general conservation laws for a multiphase medium are used [1]. Since the movement of air flows in the atmosphere is turbulent, the Reynolds equations are used to describe them [19, 43]. Taking into account the estimates, we can assume that the horizontal component of the wind speed in the forest canopy has little effect on the ignition process. For this reason, we consider the process to be quasi-one-dimensional, that is, we assume that all parameters depend on the time t and the vertical coordinate z. The formulated problem is reduced to solving the following equations [1, 27].

$$\frac{\partial \rho}{\partial t} + \frac{\partial}{\partial z}(\rho w) = \dot{m}; \tag{41}$$

$$\frac{\partial}{\partial t}(\rho w) + \frac{\partial}{\partial z}(\rho w^2) = -\frac{\partial p}{\partial z} + \frac{\partial}{\partial z}(-\rho \overline{w'^2}) - \rho s c_d w^2 - \rho g; \tag{42}$$

$$\frac{\partial}{\partial t}(\rho c_p T) + \frac{\partial}{\partial z}(\rho w c_p T) = \frac{\partial}{\partial z}(-\rho c_p \overline{w' T'}) + k(cU_R - 4\sigma T^4) + q_5 R_5 + A_S(T_s - T); \tag{43}$$

$$\frac{\partial}{\partial t}(\rho c_\alpha) + \frac{\partial}{\partial z}(\rho w c_\alpha) = \frac{\partial}{\partial z}(-\rho \overline{w' c'_\alpha}) - R_{5\alpha}, \alpha = 1, 2; \tag{44}$$

$$\frac{\partial}{\partial z}\left(\frac{c}{3k}\frac{\partial U_R}{\partial z}\right) - k(cU_R - 4\sigma T_s^4) = 0; \tag{45}$$

$$\sum_{i=1}^{4} \rho_i c_{pi} \varphi_i \frac{\partial T_s}{\partial t} = q_3 R_3 - q_2 R_2 + k(cU_R - 4\sigma T_s^4) + A_S(T - T_s); \tag{46}$$

$$\rho_1 \frac{\partial \varphi_1}{\partial t} = -R_1, \rho_2 \frac{\partial \varphi_2}{\partial t} = -R_2, \rho_3 \frac{\partial \varphi_3}{\partial t} = \alpha_c R_1 - \frac{M_c}{M_1} R_3, \rho_4 \frac{\partial \varphi_4}{\partial t} = 0; \tag{47}$$

$$\sum_{\alpha=1}^{3} c_\alpha = 1,\ p_e = \rho RT \sum_{\alpha=1}^{3} \tfrac{c_\alpha}{M_\alpha},\quad \dot{m} = (1-\alpha_c)R_1 + R_2 + \tfrac{M_c}{M_1}R_3,$$
$$R_{51} = -R_3 - \tfrac{M_1}{2M_2}R_5,\ R_{52} = \nu(1-\alpha_c)R_1 - R_5,$$
$$R_1 = k_1\rho_1\phi_1 exp\left(-\tfrac{E_1}{RT_S}\right),\ R_2 = k_2\rho_2\phi_2 T_S^{-0.5} exp\left(-\tfrac{E_2}{RT_S}\right),$$
$$R_3 = k_3\rho\phi_3 s_\sigma c_1 exp\left(-\tfrac{E_3}{RT_S}\right),$$
$$R_5 = M_2 k_5\left(\tfrac{c_1 M}{M_1}\right)^{0.25}\left(\tfrac{c_2 M}{M_2}\right)T^{-2.25} exp\left(-\tfrac{E_5}{RT}\right).$$

The initial and boundary conditions for the system of Eqs. (40) and (46) have the following form

$$t = 0 : w = 0, T = T_e, c_\alpha = c_{\alpha e}, T_s = T_e, \phi_i = \phi_{ie}; \tag{48}$$

$$z = z_0 : \frac{\partial w}{\partial z} = 0, \frac{\partial T}{\partial z} = 0, \frac{\partial c_\alpha}{\partial z} = 0, \quad -\frac{c}{3k}\frac{\partial U_R}{\partial z} = \frac{\varepsilon}{2(2-\varepsilon)}(4\sigma T_s^4 - cU_R); \tag{49}$$

$$z = h : \frac{\partial v}{\partial z} = 0, \frac{\partial w}{\partial z} = 0, \frac{\partial T}{\partial z} = 0, \frac{\partial c_\alpha}{\partial z} = 0, \quad \frac{c}{3k}\frac{\partial U_R}{\partial z} + \frac{c}{2}U_R = 2q_R(r, z). \tag{50}$$

x, *z* are the horizontal and vertical coordinates counted from the epicenter of the explosion. The terms associated with turbulent diffusion of mass, momentum, and energy are represented in terms of mean flow gradients using the Boussinesq approximation [19].

The thermodynamic, thermophysical, and structural characteristics correspond to the FCM of the pine forest canopy and were chosen according to [1]. The heat transfer coefficient A_s was chosen on the basis of data on heat exchange between the element of forest combustible material (needle, thin twig) and the environment. The initial values of the volume fractions of the phases were calculated using equations according to [1]. The value of the radiant heat flux density q_R during the flight of the Tunguska meteorite was determined using the equation for the brightness of a celestial body (36) as a result of solving problem (37)–(40).

$$q_R(r, t) = \frac{t_p I \sin L}{4\pi R_0^2},\ I = 0.5C_H\rho V^2 S_m, \tag{51}$$

where t_p is the atmospheric transmittance, L is the angle of incidence of radiation on the forest canopy, R_O is the distance from the location of the meteorite to the forest canopy. The radiation flux arriving at the upper boundary of the forest canopy after the moment of explosion $t = t_1$ was obtained using the data given in [4, 27–29]

$$q_R(r, t) = \frac{t_p P_m \sin L}{4\pi R_0^2}\begin{cases}(t - t_1)/t_m, & t < t_m \\ \exp(-k_0((t - t_1)/t_m - 1)), & t \geq t_m\end{cases} \tag{52}$$

$$t_m = t_1 + 0.032 w_0^{0.5},\ P_m = f w_0^{0.5}, \tag{53}$$

where f is an empirical coefficient that varies from 1.33 (collisional catastrophes) to 4 (technogenic explosion), the explosion power w_o was set based on the data on the explosion energy of the Tunguska celestial body, which is given in [36] and refined below.

8 The Method of Numerical Solution and Results of Calculation

The system of Eqs. (41)–(47) with initial and boundary conditions (48)–(50) for numerical integration was reduced to a discrete form using the Patankar—Spaulding control volume method [21]. Grid equations arising in the course of discretization were solved using the SIP method [22]. The velocity and pressure fields were matched iteratively using the SIMPLE algorithm [21]. The difference scheme for the system of Eqs. (41)–(47) was obtained on the basis of the modified Euler method [44]. To assess the accuracy of the schemes used and the correctness of the program, the method of a priori specified analytical solutions was used.

The sequence of solving the problem was as follows. First, at each time step, the system of ordinary differential Eqs. (37)–(40) was solved. As a result, the position of the luminous body in space above the forest phytocenosis and the value of I were determined using Eq. (36), which was used to calculate the radiation flux density $q_R(r,t)$ according to (51), reaching the forest canopy during the meteorite flight. Then the system of Eqs. (41)–(47) with initial and boundary conditions (48)–(50) was solved. Their distributions were calculated on a new time layer for given initial values of all functions. The nonlinearity of the system of algebraic equations was overcome by using the iteration method. Then the equations of chemical kinetics were solved.

Next, a transition to a new time layer was carried out, and the process was repeated in the order indicated above. Since the explosion of the meteorite, the solution (37)–(40) ceased and $q_R(r,t)$ was determined as the radiation flux released during the explosion according to Eqs. (52)–(53). The calculation process was completed when the ignition of the FCM was achieved or when it cooled down, if the radiation intensity was insufficient for ignition.

As a result of numerical integration, distributions of temperatures, mass concentrations of gas phase components, and volume fractions of solid phase components at different times at different distances from the explosion epicenter were obtained. The calculations were carried out for the total energy of the celestial body 10^{16} J [36], which is the sum of the kinetic energy of the body K_o and the energy of a possible explosion E_o. The dimensionless parameter ε_0 was used, which is determined by the equation

$$\varepsilon_0 = E_0/(E_0 + K_0) \tag{54}$$

In calculations, the value of ε_0 varied from 0.1 to 0.9. Figures 19, 20 and 21 show the temperature distributions of the gas and solid phases: (——— -, -----), mass concentrations of oxygen components and combustible pyrolysis products (—— - ------- -) and volume fractions of dry FCM (——) and water (-------) at the upper boundary of the forest canopy at different distances from the explosion epicenter for $\varepsilon_0 = 0.1$. The condition was used as the ignition criterion

$$\left.\frac{\partial^2 T}{\partial t^2}\right|_{t=t_{\acute{K}},x=x_{1*},x_2=x_{2*},z=h} = 0, \tag{55}$$

where (x_{1*}, x_{2*}, h) are the coordinates of the forest canopy point where ignition occurs. Figure 18 shows the changes in functions over time for various distances from the epicenter of the explosion ($1 - x_1, x_2 = 0$; $2 - x_1 = -10$ km, $x_2 = 0$; $3 - x_1 = -15$ km, $x_2 = 0$.).

From the analysis of the graphs (Figs. 19, 20 and 21), it can be seen that three combustion reaction modes are realized: the first, called burning, represents the actual combustion of fuel and pyrolysis products (curves 1 in Fig. 19 in the vicinity of the explosion epicenter, the second is the so-called normal ignition mode (curves 2 in Fig. 19), as well as the third non-ignition mode (curves 3 in Fig. 19).

It is implemented outside the ignition zone. In this area, drying and pyrolysis of the fuel FCM takes place, but the amount of radiant energy is not enough to ignite the FCM.

Figures 22, 23 and 24 show the temperature distributions of the gas and solid phases: (—— - $\overline{T}$, -----$\overline{T}_S$,), mass concentrations of oxygen components and combustible pyrolysis products (—— - $\overline{c}_1$,------ - $\overline{c}_2$) and volume fractions of dry

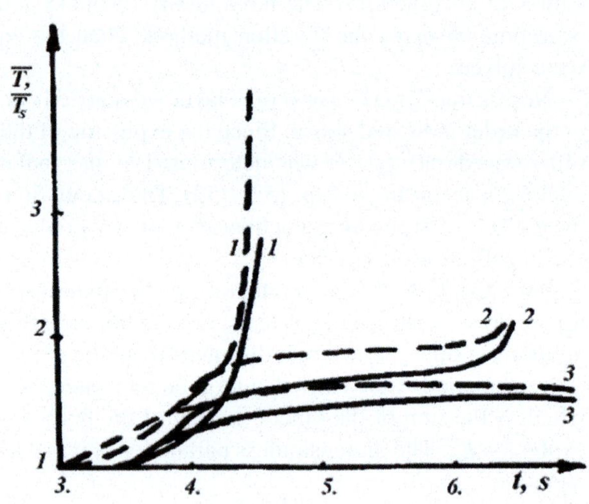

Fig. 19 The temperatures of the gaseous (——) and condensed (------) phases over time at the upper boundary of the forest canopy for various distances from the epicenter

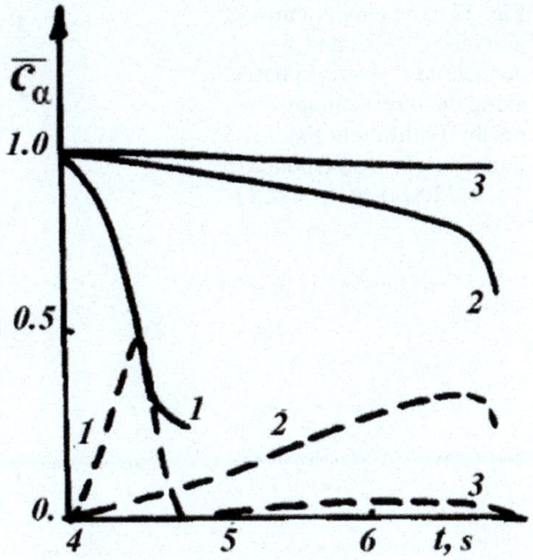

Fig. 20 The mass concentrations of oxygen (——) and pyrolysis products (------) over time at the upper boundary of the forest canopy for various distances from the epicenter

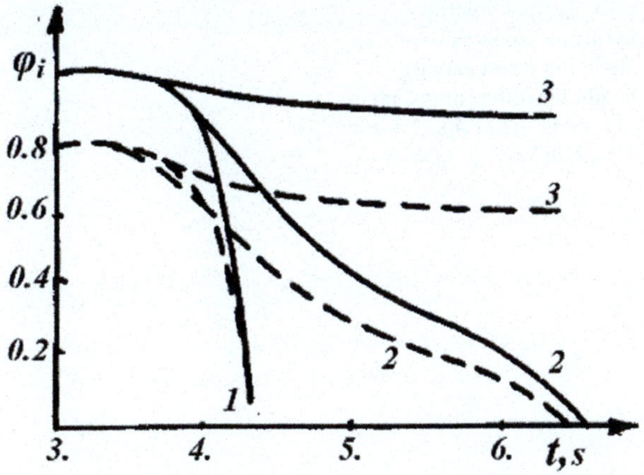

Fig. 21 The volume fractions of the condensed phase (——, ------) over time at the upper boundary of the forest canopy for various distances from the epicenter

FCM (—— $\overline{\varphi}_1$,) and water (------$\overline{\varphi}_2$) according to the height of the forest canopy for different time points (1 – 4 s, 2 – 4.5 s.), for $x_1 = 0$, $x_2 = 0$ and $\varepsilon_0 = 0.1$. It can be seen from the graphs that, at first, an inert heating of the FCM occurs (the temperature of the solid phase rises under the influence of the radiant flux), and then the gas phase heats up from the solid phase. Then there is the evaporation of water and the pyrolysis of FCM with the release of volatile combustible products. The latter then ignite at the time t_*, when the pyrolysis of FCM is still ongoing.

Thus, in accordance with [1, 11], gas-phase ignition of gaseous pyrolysis products takes place. As a result of a large number of calculations for various distances and directions from the epicenter of the explosion, it was found that the ignition source has an oval shape. Also, it can be approximated in the form of two semi-ellipses

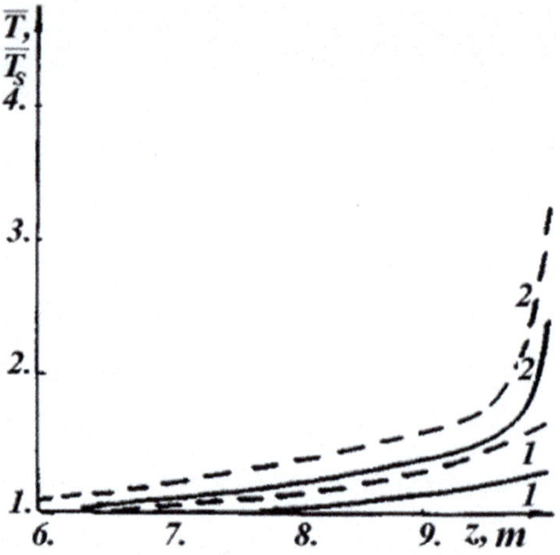

Fig. 22 The temperatures of gaseous (——$\overline{T}$) and condensed (------$\overline{T}_S$) phases along the forest canopy height for different time moments (1—corresponds to 4 s, 2—4.5 s.), at $x_1 = 0, x_2 = 0$

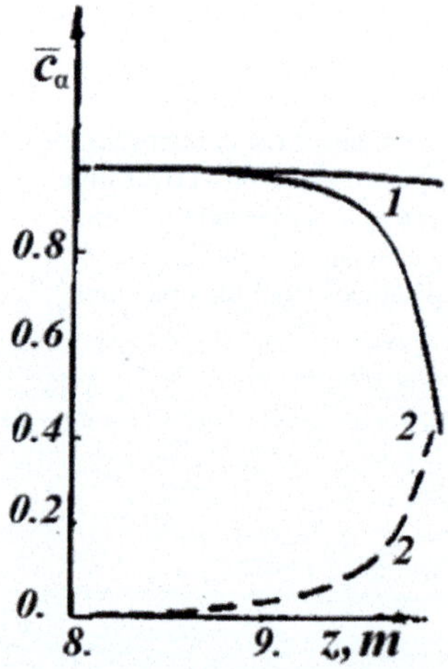

Fig. 23 The mass concentrations of oxygen (——$\overline{c}_1$) and volatile pyrolysis products (----- $\overline{c}_2$) along the forest canopy height for different instants of time (1 – t = 4 s, 2 – t = 4.5 s.), at $x_1 = 0, x_2 = 0$

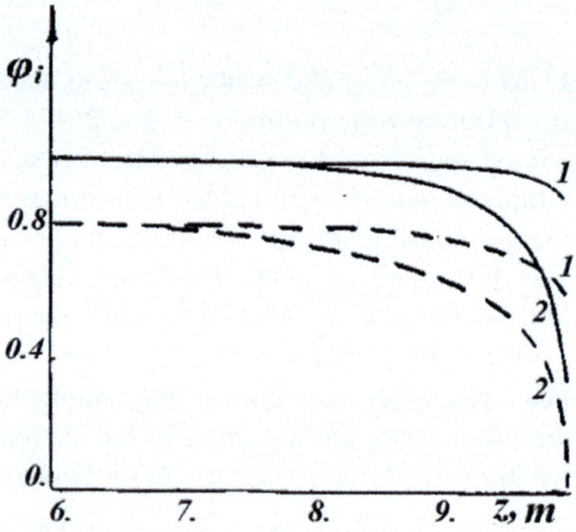

Fig. 24 The volume fractions of condensed phase (——$\overline{\varphi}_1$) and (------$\overline{\varphi}_2$) along the forest canopy height for different instants of time (1 – t = 4 s, 2 – t = 4.5 s.), at $x_1 = 0, x_2 = 0$

docked to each other, having the same minor semi-axis b and different major semi-axes a_1 and a_2 (corresponding to the directions from the arrival side of the celestial body and the direction behind the epicenter of the explosion). Calculations show that the shape and size of the ignition zone are significantly affected by the value of ε_0 and parameters characterizing various tree species (Tables 1, 2 and 3).

Table 1 shows the values of the semi-major and minor semi axes in the elliptical part of the ignition zone in km for the corresponding values of the ignition time t_* in seconds for $\varepsilon_0 = 0.1$. Here and below, the values of a_1, a_2, b, and t_* were obtained at $f = 1.33$.

Table 3 presents the same values as in the previous tables, but for $\varepsilon_0 = 0.5$.

It can be seen that the obtained agreement with the experimental data from [36] takes place at $f = 1.33$, $\varepsilon_0 = 0.1$ for a pine forest. The closer ε_0 to 1, the more the contours of the ignition zone resemble a circle. This is explained by the fact that the share of energy emitted during the flight of a celestial explosion decreases and the amount of radiant energy released during the explosion increases. Since during the explosion of a meteorite the radiant flux acts symmetrically on the forest canopy, the ignition zone with an increase in ε_0 approaches the shape of a circle, as is the case in the case of a nuclear explosion. The following data were used in the calculations: for a pine forest $k_1 = 3.2$, $E_1/R = 4402$ K, $\alpha_c = 0.33$, $k_2 = 1.88$, $E_2/R = 2720$ K, $W =$

Table 1 The values of the maximum sizes of ignition zones for various tree species at $\varepsilon_0 = 0.1$

	Larch			Pine			Birch		
a_1, km	0	8.5	0	0	11	0	0	6	0
a_2, km	0	0	6.5	0	0	8	0	0	4.5
b, km	7	0	0	6	0	0	5	0	0
t_*, s	7.3	7.0	7.0	6.6	7.2	7.0	8.0	7.8	7.0

Table 2 The values of the maximum sizes of ignition zones for various tree species at $\varepsilon_0 = 0.3$

	Larch			Pine			Birch		
a_1, km	0	12	0	0	12.5	0	0	9.5	0
a_2, km	0	0	11	0	0	12	0	0	8.5
b, km	11	0	0	12	0	0	8.5	0	0
t_*, s	8.9	8.7	8.9	7.0	7.6	7.4	8.0	8.2	8.0

Table 3 The values of the maximum sizes of ignition zones for various tree species at $\varepsilon_0 = 0.5$

	Larch			Pine			Birch		
a_1, km	0	13	0	0	15.5	0	0	11	0
a_2, km	0	0	13	0	0	15	0	0	11
b, km	13	0	0	15	0	0	11	0	0
t_*, s	8.6	8.4	8.6	8.0	7.5	8.0	8.8	8.4	8.8

0.8; for larch $k_1 = 3.2$, $E_1/R = 4402$ K, $\alpha_c = 0.33$, $k_2 = 3.2 \times 10^4$, $E_2/R = 5293$ K, $W = 1$; for birch $k_1 = 0.06$, $E_1/R = 2310$ K, $\alpha_c = 0.17$, $k_2 = 4.4 \times 10^5$, $E_2/R = 6711$ K, $W = 1.7$, which were obtained at a heating rate of FCM of 0.0966 K/s [1]. The values of the moisture content are typical for the forest fuel for the summer season, when the Tunguska meteorite fell. As a result of the calculations, the contours of the ignition zone under the influence of radiation from the Tunguska meteorite were obtained, which are an asymmetric ellipse, elongated in the direction of the arrival of the body. Figures 25, 26 and 27 in the form of closed curves show the contours of the ignition zones, which are realized at different times from the moment the Tunguska celestial body enters the surface evil atmosphere for various tree species (pine, larch and birch, respectively).

The resulting forms of combustion centers are explained by the fact that before the moment of explosion, as a result of the action of a radiant heat flux from the heated Tunguska meteorite, heating and drying of the FCM take place in the zone adjacent in the direction of flight. As a result, in these areas, ignition after the meteorite explosion

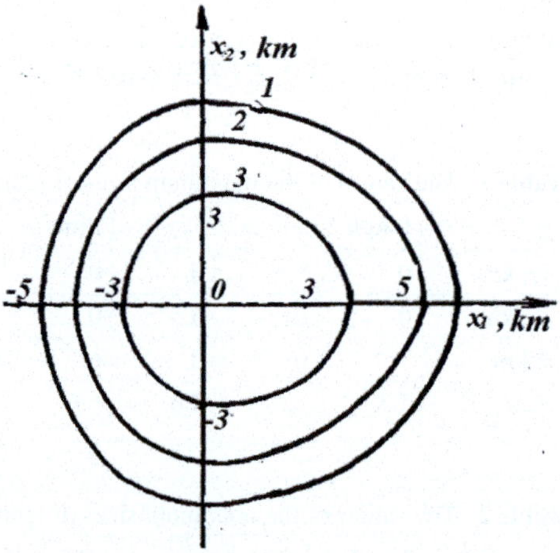

Fig. 25 The dynamics of the development of the ignition zone for the pine forest: 1 – $t = 7.0$ s, 2 – $t = 5.0$ s, 3 – $t = 4.3$ s

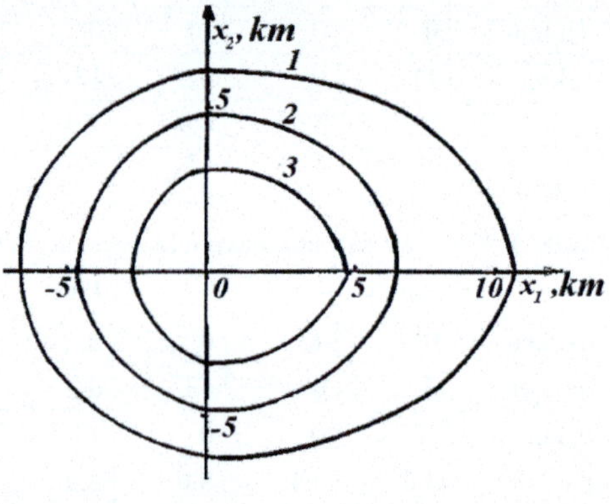

Fig. 26 The dynamics of the development of the ignition zone for the pine forest: 1 – $t = 7.0$ s, 2 – $t = 5.0$ s, 3 – $t = 4.3$ s

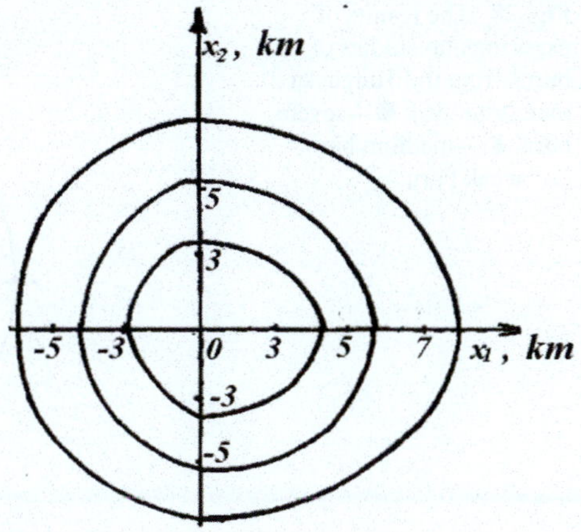

Fig. 27 The dynamics of the development of the ignition zone for the pine forest: 1 – $t = 7.0$ s, 2 – $t = 5.0$ s, 3 – $t = 4.3$ s

is realized faster than in the right neighborhood of the explosion epicenter. Moreover, the ignition zone obtained for the pine forest has the maximum dimensions. The FCM temperature at the time of ignition (T_*—ignition temperature), which corresponds to the values $t = t_*$, where t_*, is the ignition time, is not a constant value, but depends on the coordinates of the contour points and the type of vegetation. This dependence is weak. The spread of ignition temperature values depends on the type of vegetation and varies from one circuit to another within 600 K $\leq T_* \leq$ 700 K.

The shape of the ignition contours and their sizes at $\varepsilon_0 = 0.1$ agree qualitatively with the observational data in the explosion zone of the Tunguska celestial body [36]. Figure 28 presents the results of experimental studies on the study of the ignition zone and forest fire caused by the explosion of the Tunguska meteorite [36]. The hatched circles indicate the places of severe burns of trees as a result of the action of light radiation, half-shaded circles—moderate burns and white circles—weak burns. Due to the fact that in numerical calculations the direction of motion of the celestial body was considered from left to right, the elongated part of the ellipse.

It follows from the analysis of the results that the ignition circuits for natural collision and technogenic are different. In the first case, the contour is composed of *a* circular arc in the vicinity of the epicenter and an elliptic curve away from the epicenter. In the second case, the contours at any time are circles. The difference between these contours is maximum for small ε_0, and for $\varepsilon_0 \to 1$, the contours of the first type become circles for any $C_н$. As f grows, the values of a and b increase, while t_* decreases. This is explained by an increase in the density of the radiant heat flux. Since the resulting gaseous combustible pyrolysis product of forest fuel does not completely burn out when igniting forest combustible materials, a cloud of gaseous combustible pyrolysis products mixed with air appears in the forest canopy and above it in the ignition zone. This cloud can explode under the action of ballistic and explosive shock waves generated during the flight and explosion of a meteorite. In particular, such a situation can take place if the inequality

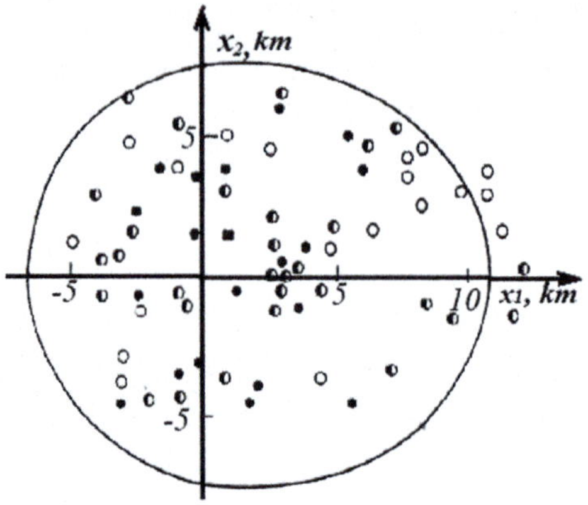

Fig. 28 The results of experimental studies of tree burns from the Tunguska meteorite [36]: ●—severe burn, ◐—medium burn, ○—weak burn

$$t_* \geq t_V, \text{ where } t_V = \int\limits_0^{H_*} \frac{dz}{D_z} = \frac{H_*}{D_{z*}} \tag{56}$$

where t_V is the time of arrival of the shock wave to the earth's surface, H_* is the height of the explosion, D_z is the projection of the velocity of propagation of the shock wave on the z axis, $D_{z*} = D_z(z_*)$, where $0 < z_* < H_*$ is the value of z according to the theorem about the average. Estimates show that $t_V \sim$ 3–6 s, i.e. Inequality (56) is satisfied and the shock wave can initiate a mixture of formed gaseous substances of forest fuels with air, as a result of which the destructive effect of the incident shock wave is enhanced. In this regard, it is of interest to estimate the total mass of gaseous combustible products of pyrolysis in the surface layer of the atmosphere $M(t)$.

By definition

$$M(t) = \int_0^h \int_{S_*} \rho c_2 dxdydz + \int_0^{t_*} \int_{S_*} \rho w c_2 dxdydt, \tag{57}$$

where S_* is the maximum value of the ignition zone area. The first term on the right side of (57) characterizes the mass of gaseous pyrolysis products formed in the ground cover and forest canopy, and the second term characterizes the amount of pyrolysis products that are carried into the surface layer of the atmosphere by the ascending gas flow. Figure 29 shows graphs of $M(t)$ for various values of ε_0. It can be seen that $M(t)$ increases with time and ε_0. It is explained by the fact that with an increase in ε_0, the amount of heat released as a result of the glow of a celestial body during its flight increases, which affects the FCM of the forest canopy and, consequently, more volatile pyrolysis products are released. As estimates show, the maximum mass of products at $t = t_*$ and $0.1 \leq \varepsilon_0 \leq 0.9$ and $C_н = 0.1$ varies within $4 \times 10^7 < M < 5 \times 10^8$ kg. The main amount of gaseous combustible products of FCM pyrolysis is CO.

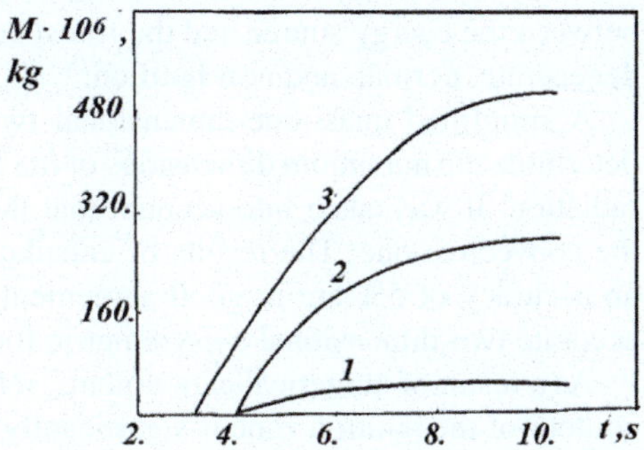

Fig. 29 Change in the mass of combustible pyrolysis products in the ignition zone; $1 - \varepsilon_O = 0.1$, $2 - \varepsilon_O = 0.5$, $3 - \varepsilon_O = 0.9$

Therefore, bearing in mind that during the oxidation of 1 kg of CO, energy $q = 10^7$ J is released, we obtain that the TNT equivalent of an explosion of gaseous pyrolysis products varies for the Tunguska celestial body in the range from 100 kilotons to 1.2 megatons. In particular, at $E = 10^{16}$ J, $\varepsilon_0 = 0.1$, the TNT equivalent of gaseous pyrolysis products is 2×10^{14} J, which corresponds to 20% of the explosion energy of the Tunguska meteorite ($Eo = 10^{15}$ J), which was refined by comparing the calculated ignition circuits with the data [36, 45].

It would be said that the above estimates do not exhaust the analysis of the problem of the explosion of gaseous products of the pyrolysis of forest fuels. It is possible that their initiation occurs as a result of the action of a shock wave reflected from the Earth's surface rather than an incident one. In other words, the mechanism for initiating the detonation of a fuel-air cloud deserves a separate study.

9 Conclusion

Thus, in this work, mathematical modeling of the occurrence of massive forest fires under the action of light radiation as a result of man-made and natural disasters was carried out, which allows us to draw the following conclusions.

Based on the general mathematical model of the theory of forest fires, new mathematical models have been developed for the problems of ignition of forest tracts from light radiation as a result of man-made and collisional disasters with celestial bodies such as the Tunguska meteorite. The turbulence of the flow, the two-temperature environment and the main physical and chemical processes (drying and pyrolysis of forest combustible materials, chemical reactions of combustion of gaseous and afterburning of condensed pyrolysis products) are taken into account.

New physical–mathematical formulations of the problem of ignition of forest tracts as a result of man-made and natural (collisional) disasters are given, taking into account the effect of the light radiation of a celestial body on the forest canopy during its flight in the atmosphere.

Mathematical modeling of the process of ignition of forest tracts under the influence of light radiation was carried out, which showed that, depending on the distance between the energy source and the forest canopy, three ignition modes are realized: degenerate, normal, and non-ignition.

A simplified quasi-one-dimensional two-temperature statement is proposed to determine the maximum dimensions of the zone of ignition of forest tracts from light radiation. It was taken into account that the radiant heat flux significantly exceeds the convective one. The results of calculations of the sizes of ignition zones with an accuracy of 5% are in good agreement with the results obtained using a more accurate two-dimensional axisymmetric formulation.

As a result of numerical experiments, it has been established that the shape of the contour of the ignition zone is significantly affected by the trajectory of the flight of a celestial body and the fraction of the kinetic energy of the oncoming flow, which turns into light radiation C_H, and the area of the ignition zone depends on the total energy of the celestial body E, C_H and on the ratio of the explosion energy to the total energy ε_0. In particular, it is shown that the shape of the ignition zone for the Tunguska meteorite can be represented as an oval elongated along the projection of the celestial body arrival trajectory onto the underlying surface. It is shown that within the framework of the proposed mathematical model, it is possible to obtain not only qualitative, but also quantitative agreement between the shape and dimensions of the contour with the known observational data at $E = 10^{16}$ J, $E_o = 10^{15}$ J and $C_H = 0.1$.

As a result of mathematical modeling, it was found that for various forest stands the sizes of ignition zones, other things being equal, are arranged in descending order as follows: pine, larch and birch forests. The minimum ignition time is realized for pine stands, and the maximum for birch stands. Based on the results of numerical calculations of physical and chemical processes in the ignition zone from the Tunguska meteorite, an assessment was made of the intensification of the shock wave effect on the forest. It has been established that up to 20% of the recorded explosion energy can be provided as a result of the detonation of volatile combustible pyrolysis products, which are formed in the forest canopy by the time the shock wave arrives. Thus, when estimating the total energy of the explosion of the Tunguska meteorite from the action of a shock wave on a forest, it is also necessary to take into account its interaction with the gaseous products of the pyrolysis of forest combustible materials.

References

1. Grishin, A.M.: Mathematical Modeling of Forest Fires and New Methods of Fighting Them. Publishing House of the Tomsk University, Tomsk (1997)
2. Pittock, T.P., Ackerman, P.J., Crutzen, Michael C., MacCracken, C.S., Shapiro, R.P.: Turco: environmental consequences of nuclear war, physical and atmospheric effects. Wiley (1989)
3. Bronshten, V.A.: Physics of Meteoric Phenomena. D. Reidel Publishing Company, Dordrecht, Holland (1983)
4. Glasstone, S., Dolan P.J.: The Effects of Nuclear Weapons. U.S. Department of Defense and the U.S. Department of Energy (1977)

5. Gostintsev, Y.A., Makhviladze, G.M., Novozhilov, V.B.: The initial stage of development of a large fire initiated by radiation//combustion of condensed systems. In: Proceedings of the IX All-Union Symposium on Combustion and Explosion. Chernogolovka: IHF (1989)
6. Ryzhov, A.M., Gostintsev, Yu.A.: Formation of a large fire caused by radiation. Fire Sci. Technol. **16**, 13–28 (1996)
7. Grishin, A.M.: Mathematical Models of Forest Fires. TSU Publishing House, Tomsk (1981)
8. Perminov, V.A.: Mathematical modelling of wildland fires initiation and spread using a coupled atmosphere-forest fire setting. Chem. Eng. Trans. **70**, 1747–1752 (2018)
9. Grishin, A.M., Perminov, V.A.: Ignition of forest tracts by a high-altitude source of radiant energy. Combust. Explos. Shock Waves **32**, 565–571 (1996)
10. Gostintsev, Y.A., Ryzhov, A.M.: Numerical simulation of the dynamics of free-burning flames, fire whirls, and firestorms. Fluid Dyn. **29**, 780–788 (1994)
11. Perminov, V.A.: Mathematical modeling of the occurrence of mass and crown forest fires, taking into account the radiation-convective heat and mass transfer and the two-temperature environment. Ph.D thesis. Tomsk State University, Tomsk (1995)
12. Grishin, A.M., Perminov, V.A.: Transition of the forest ground fire to crown fire. Combust. Explos. Shock Waves **26**(6), 644–651 (1991)
13. Strelets, M.: Turbulence modelling in convective flow of fires. In: Proceedings of the 4th International Seminar on Fire and Explosion Hazards, North Ireland, Londonderry (2003)
14. Perminov, V.A.: Mathematical modeling of crown forest fire initiation. Lect. Notes Comput. Sci. **2667**, 549–557 (2003)
15. Dubov, A.S., Bykova, L.P., Morunich, S.V.: Turbulence in Vegetation. Gidrometeoizdat, Leninrad (1978)
16. Grishin, A.M., Perminov, V.A.: The influence of complex radiative-convective heat transfer on the transition of a surface fire to a crown fire. In: Heat and Mass Transfer—International Forum. Radiation and combined heat transfer (Minsk: ITMO), pp. 47–49 (1988)
17. Fomin, A.A.: The structure of the flow and the heating of the environment over the local focus of a forest fire. Ph.D thesis. Tomsk State University, Tomsk (1989)
18. Perminov, V.A., Goudov, A.M.: Mathematical modeling of forest fires initiations, spread and impact on environment. Int. J. Geomate **13**, 93–99 (2017)
19. Sedov, L.I.: Mechanics of Continuous Media, vol. 1. World Scientific (1997)
20. Grishin, A.M., Perminov, V.A.: Mathematical modeling of the ignition of tree crowns. Combust. Explos. Shock Waves **34**, 378–386 (1998)
21. Patankar, S.V.: Numerical Heat Transfer and Fluid Flow. Hemisphere Publishing Corporation (1981)
22. Stone, H.L.: Iterative solution of implicit approximations of multidimensional partial differential equations. SIAM J. Numer. Anal. **5**, 530–558 (1968)
23. Altamirova, E., Perminov, V.A.: Mathematical modeling of forest canopy ignition by thermal radiation from a hydrocarbon explosion. Combust. Explos. Shock Waves **55**, 574–579 (2019)
24. Johnstona, C.O., Stern, E.C.: A model for thermal radiation from the Tunguska airburst. Icarus **327**, 48–59 (2019)
25. Agranat, V., Perminov, V.: Mathematical modeling of wildland fire initiation and spread. Environ. Modell. Softw. **125** (2020)
26. Grishin, A.M., Filkov, A.I.: A deterministic-probabilistic system for predicting forest fire hazard. Fire Saf. J. **46**, 56–62 (2011)
27. Grishin, A.M., Perminov, V.A.: Mathematical modeling of the ignition of the forest canopy from the Tunguska meteorite. Izvestiya SO RAS. Sib. J. Phys. Technol. **6**, 112–117 (1992)
28. Grishin, A.M., Permnov, V.A.: On the ignition of forest areas as a result of the Tunguska meteorite explosion. Combust. Explos. Shock Waves **29**(6), 668–673 (1993)
29. Nikolaev, Y.A., Fomin, P.A.: Identification of the Tunguska event with the explosion of a methane-clouds initiate by a slow- and flat-moving small meteorite. Combust. Explos. Shock Waves **47**(2), 226–240 (2011)
30. Grishin, A.M., Efimov, K.N., Perminov, V.A.: Mathematical modeling of the occurrence of massive forest fires. In: Proceedings of the International Conference: "Forest Fires: occurrence, Spread and Environmental Consequences", pp. 43–44 (1995)

31. Kurbatsky, N.P.: About the forest fire in the area of the Tunguska fall in 1908. Meteoritika **24**, 53–57 (1976)
32. Vasiliev, N.V., Lvov, Y.A.: Radiant burn of trees in the area of the fall of the Tunguska meteorite. Priroda **3**, 22–23 (1974)
33. Zenkin, G.M., Ilyin, A.G.: On the radiation burn of trees in the area of the explosion of the Tunguska meteorite. Meteoritika **24**, 129–140 (1964)
34. German, B.R.: Physics of the 1908 Tunguska Event. Verlag, Freiburg (2022)
35. Fast, V.G.: Fall of the forest produced by the Tunguska meteorite. In: Proceeding of the MEETING CURRENT STATE of the Problem of the Tunguska Meteorite, pp. 41–42. Publishing House of Tomsk State University (1971)
36. Korobeinikov, V.P., Chushkin, P.I., Shurshalov, L.V.: Complex modeling of flight and explosion in the atmosphere of a meteoroid. Astrono. Bull. **25**(3), 327–343 (1991)
37. Perminov, V.A., Soprunenko, E.E.: An initiation of forest fires as a result of gas pipeline accidents. Int. J. Circ. Syst. Signal **12**, 210–214 (2018)
38. Grishin, A.M., Zverev, V.G., Gruzin, A.D.: Mathematical theory of crown forest fire. Heat Transf. Sov. Res. **17**(6) (1986)
39. Stulov, V.P.: Gas dynamical model of the Tunguska fall. Planet. Space Sci. **46**(2), 253–260 (1998)
40. Kang, S.W., Reitter T.A., Takata A.N.: Analysis of large urban fires. AIAA Paper 85-0457, AIAA 23-d Aerospace sciences meeting (1985)
41. Siegel, R., Howell, J.R.: Thermal Radiation Heat Transfer. Hemisphere Publishing Corp, New York (1981)
42. Ozicik, M.N.: Radiative Transfer and Interactions with Conduction and Convection. J. Wiley, New York (1973)
43. Monin, A.S., Yaglom, A.M.: Statistical Fluid Mechanics, vol. 1. MIT Press, Cambridge (1975)
44. Roache, P.J.: Computational Fluid Dynamic. Hermosa Publishers (1976)
45. Robertsona, D.K., Mathias, D.L.: Hydrocode simulations of asteroid airbursts and constraints for Tunguska. Icarus **327**(15), 36–47 (2019)

Optical Methods for Detecting Local Microdefects in Cable Products

Ivan Razuvaev, Evgeny Fedorov, and Vitaly Redko

Abstract The paper addressed the study of feasibility of optical methods to measure the diameter of extended products to perform in-process detection of local microdefects in cable products. The analysis of data on through-transmission methods and the power measurement method showed that the power measurement method is most appropriate for implementation of the microdefect detection device since it enables measuring at the desired frequency. The experimental study confirmed the efficiency of the power measurement method, and revealed a significant drawback of the method, namely, an increased error due to the inhomogeneous distribution of optical flow power when the test object moves through the measurement zone.

1 Introduction

The cable industry is one of the highly crucial sectors in the modern world. Despite its high manufacturability, energy intensity, resource intensity and a high degree of automation, this industry keeps a rapid pace of advancement.

Detection of local defects and measurement of their geometric parameters are the most complex processes in the cable industry [1–5]. This inspection is highly relevant due to a significant effect of local insulation defects and current-carrying conductors on both the strength and electrical characteristics of products [1, 6–9].

Non-destructive testing devices most widely used in production employ the electric spark method. A significant drawback of these devices is that they only allow

I. Razuvaev (✉) · E. Fedorov
School of Non-Destructive Testing, National Research Tomsk Polytechnic University, 30 Lenin Ave., Tomsk 634050, Russia
e-mail: razuvaev_ivan@mail.ru

E. Fedorov
e-mail: evgeny_fyodorov@list.ru

V. Redko
Scientific-Production Enterprise "Redwill", 7 Savinykh Str, Tomsk 634028, Russia
e-mail: Ultratone@rambler.ru

E. Lysenko et al. (eds.), *Emerging Trends in Materials Research and Manufacturing Processes*, Engineering Materials, https://doi.org/10.1007/978-3-031-38964-1_3

detection of through defects and thinning in the insulation of current-carrying conductors, and therefore new measuring methods and devices are required to improve the process of inspection [10, 11].

Optical inspection methods are most appropriate to detect local defects, since, in contrast to X-ray and capacitance methods, they provide high resolution, speed, and measurement frequency at a relatively low cost of devices [12, 13].

2 Theoretical Part

The currently faced problem is detection of local defects with a length of 0.5 mm and a height of 10 μm at a production line speed of up to 3000 m/min. The power measurement method and through-transmission methods are most appropriate to meet the given requirements.

2.1 *Power Measurement Method*

The power measurement method is based on the dependence of the power of the parallel light beam on its cross-sectional area (Fig. 1). A parallel light flux generated using an optical radiation source and a converging lens passes through the inspection zone, where the test object is located, which prevents propagation of part of the radiation, thereby reducing the size of the flux and, hence, its power [14]. Power measurement is performed using a light detector; a converging lens focuses the light flux incident on the detector.

The power measurement method is most promising for flaw detection in cable products due to the low cost of the finished device and the highest measurement frequency, which is critical since production lines operate at high speed [15–17]. Devices based on this method are difficult to implement, as the distribution of optical radiation power is inhomogeneous due to the features of currently used light sources and aberrations of the optical system, which significantly increases the measurement error [18].

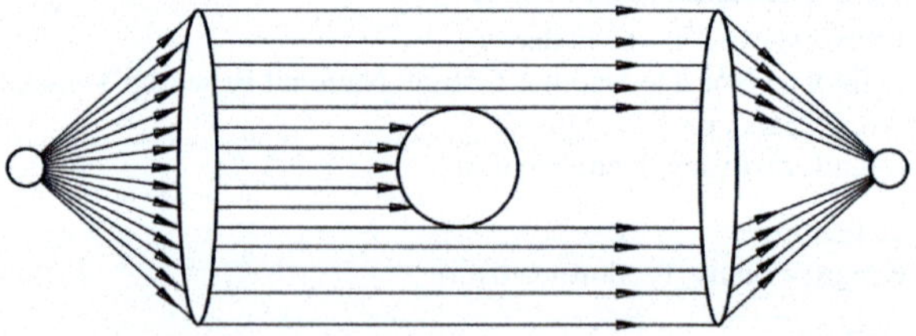

Fig. 1 Radiation power measurement method

2.2 Through-Transmission Methods

Through-transmission methods imply measurement of the diameter using matrix photocells by measuring the size of the shadow cast by the test object [19, 20]. Measurement by through-transmission methods is performed using both divergent and parallel light beams.

The through-transmission method implemented in a diverging beam (Fig. 2) is most widely used in devices designed to measure the diameter of extended products due to its high stability and measurement accuracy, as well as design simplicity [21]. The disadvantage of using the divergent flux to detect local defects is the need for expensive high-performance computing devices to calculate the diameter from the size of the shadow and the distance between the object and the photosensitive matrix. The position of the object is determined using an additional measuring axis shifted by a given angle [22].

The use of a parallel light flux (Fig. 3) is limited by the aberrations of the optical system used to generate a parallel beam that have a significant negative impact on the measurement error [23].

The basic reason that hampers the use of through-transmission methods to detect local defects is the low measurement frequency caused by the frequency limitations of matrix photocells used to measure the cast shadow. Limited measurement frequency in practice suggests the options to choose: high production line speed and increased minimum detectable defect length or reduced production speed, but preserved required metrological characteristics [24, 25].

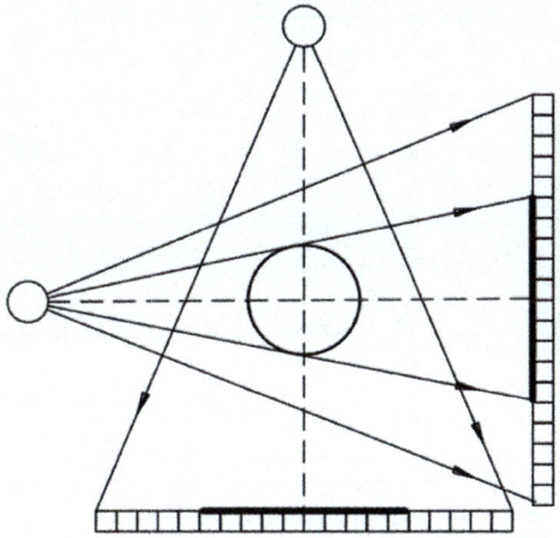

Fig. 2 Through-transmission method implemented in a divergent beam

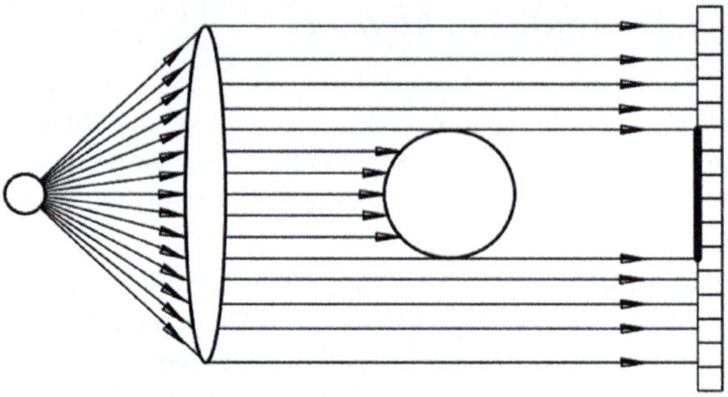

Fig. 3 Through-transmission method implemented in a parallel beam

2.3 Evaluation of the Effect of Blind Zones

The main limitation of using optical methods for inspecting microdefects in cable products is the presence of blind zones located in front and behind the test object relative to the light propagation direction. A defect located in the blind zone can be skipped, which is unacceptable [26]. This problem can be solved by adding measuring axes that are shifted by a given angle relative to the axis of symmetry of the test object and overlap blind zones (Fig. 4).

To evaluate the effect of blind zones, the probability of location of bump type defects with height h in the inspection zone was determined using Equation

$$\delta(h) = \frac{360^\circ - (\alpha(h) \cdot n)}{360^\circ} \cdot 100\%,$$

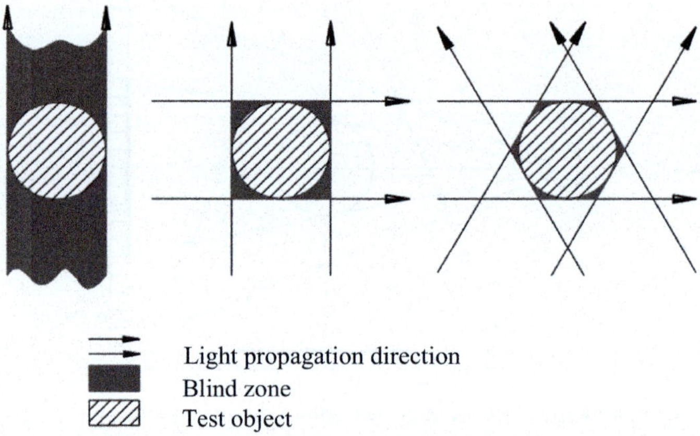

Fig. 4 Blind zones of one-axis, two-axis and three-axis devices

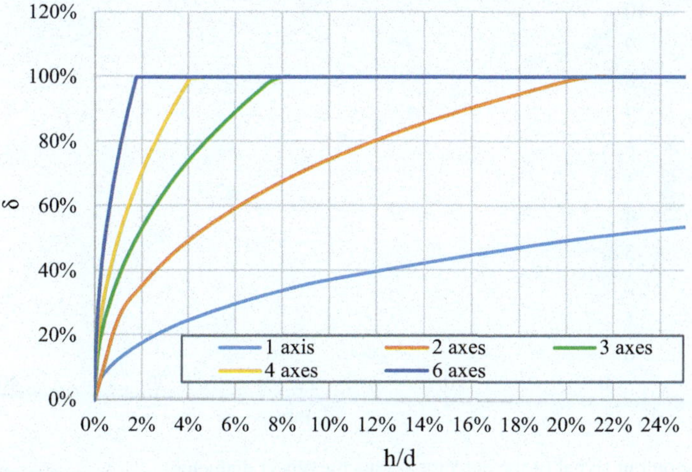

Fig. 5 Probability of the defect location in the inspection zone

where $\alpha(h)$ is the angle between two extreme positions of the defect with height h in a blind zone, and n is the number of blind zones.

The calculation results plotted in Fig. 5 show that the increased number of measuring axes decreases the maximum height of the defect, which may be skipped by the measuring device.

3 Experimental Part

3.1 Study of the Effect of the Object Diameter

To evaluate the power measurement method feasibility for local defect detection, we studied the effect of the test object diameter on the output voltage of the light detector. Round gauges located in the center of the inspection zone were a test object.

An SPL PL90_3 pulsed diode laser was used as the source of optical radiation; a SFH2701 PIN photodiode was used as a detector. Converging lenses had a length of 55 mm and an optical power of 6 diopters.

Figure 6 illustrates the dependence of the output voltage of the detector on the object diameter. As can be seen from the graph, the dependence slightly deviates from a linear function. The deviation is likely due to inhomogeneous distribution of optical radiation power in the inspection zone.

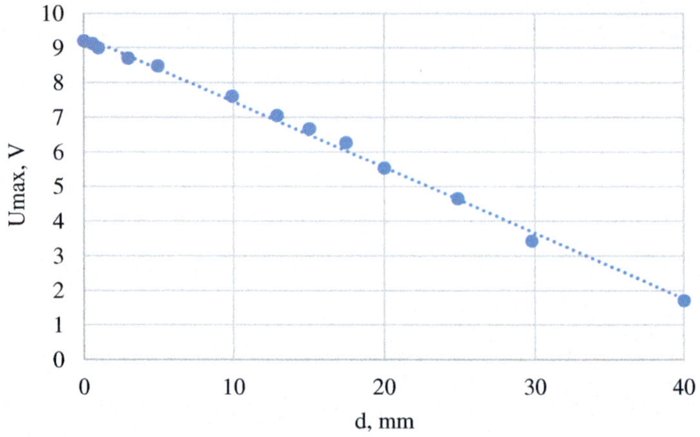

Fig. 6 The output signal of the detector versus the object diameter

3.2 Evaluation of the Effect of Luminous Flux Distribution

Based on the dependences of the output voltage of the light detector on the position of the test object in the inspection zone, luminous flux distribution was assessed using two round gauges 5 and 10 mm. Figure 7 shows the calculated and experimental dependences of the output voltage on the position of the test object. Theoretical dependencies were calculated based on the data on distribution of the radiation power of the laser used. The experimental data deviate from the theoretical ones due the optical system aberrations. The maximum deviation was 5%, which corresponds to a change in the diameter by 250 μm.

3.3 Operation Speed

To assess the performance rate, the model of the flaw detector measuring unit was connected to a personal computer via a data acquisition board. Due to hardware limitations of the equipment used, the measurement frequency was set to 10 kHz.

An 8-mm-diameter cable with artificially created defects was used as a test object (Fig. 8). The speed of cable movement was 3.5 m/min.

The graph presented in Fig. 9 shows that the developed measuring unit, based on the power measurement method, enables scanning extended products for local defects. A measurement frequency not less than 100 kHz is required to measure the diameter with a step of 0.5 mm at a line speed of 3000 m/min with regard to the data obtained.

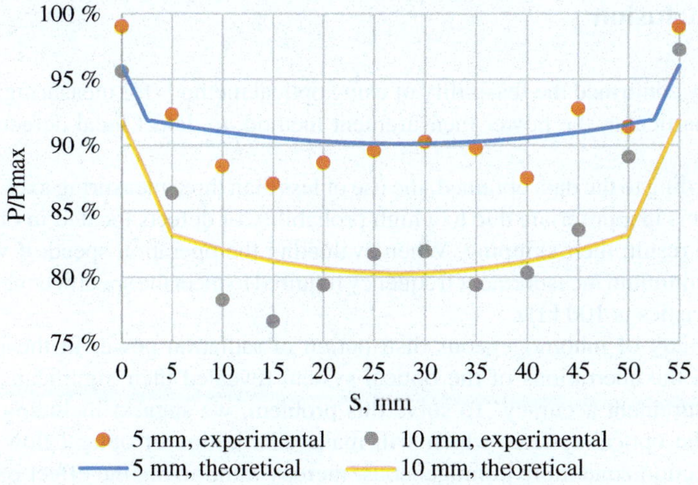

Fig. 7 Luminous flux versus the object position in the inspection zone

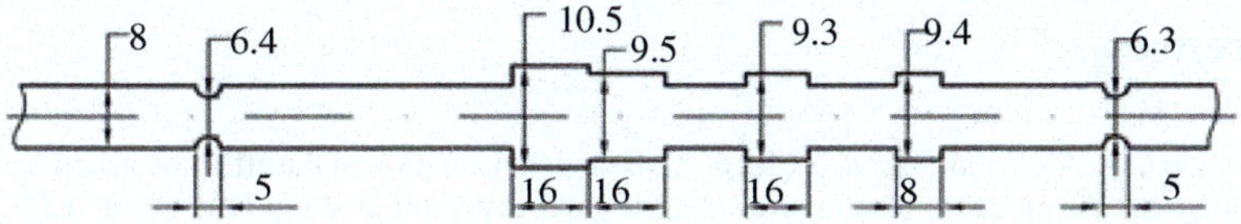

Fig. 8 Test object

Fig. 9 Results of the test object inspection

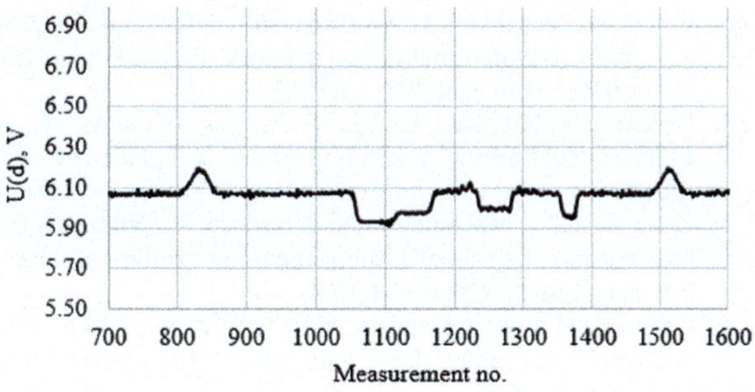

In addition, the graph shows that the measuring unit smoothes a stepwise change in the diameter. This is due to the fact that measurement is performed not for the diameter at a specific point of the object, but for the average value of the diameter in the zone equal to the width of the inspection zone.

4 Conclusion

The study confirmed the feasibility of using optical methods for measuring the diameter, in particular, the power measurement method, to detect local defects in cable products.

According to the data obtained, the use of less than three measuring axes for defect detection is inappropriate due to a high probability of defects located in blind zones and, as a result, their skipping. When evaluating the operation speed; it was found that the minimum measurement frequency required to meet the specified metrological characteristics is 100 kHz.

The study of inhomogeneous distribution of radiation power in the inspection zone and the aberrations of the optical system revealed their significant effect on the measurement accuracy. To solve this problem, we suggest including a Powell lens in the optical system, which will make distribution of optical flow power in the inspection zone more homogeneous, thereby minimizing the effect of the main source of error.

References

1. Alboyaci, B., Aytac, M., Berat, Y., Ince, A.: Elsevier evaluation of the effect of structural defects in the heat-shrink cable terminal on electric field distribution. Eng. Fail. Anal. **132**, 105920 (2022)
2. Benidir, A., Flamand, O., Gaillet, L., Dimitriadis, G.: Impact of roughness and circularity-defect on bridges stability. J. Wind Eng. Ind. Aerodyn. **137**, 1–13 (2015). https://doi.org/10.1016/j.jweia.2014.11.010
3. Uckol, H., SuatIlhan, A.: Workmanship defect classification in medium voltage cable terminations with convolutional neural network. Electr. Power Syst. Res. **194**, 107105 (2021). https://doi.org/10.1016/j.epsr.2021.107105
4. Lysenko, E., Nikolaev, E., Vlasov, V., Surzhikov, A.: Microstructure and reactivity of Fe2O3–Li2CO3–ZnO ferrite system ball-milled in a planetary mill. Thermochim. Acta **664**, 100–107 (2018)
5. Lysenko, E.N., Malyshev, A.V., Vlasov, V.A., Nikolaev, E.V., Surzhikov, A.P.: Microstructure and thermal analysis of lithium ferrite pre-milled in a high-energy ball mill. J. Therm. Anal. Calorim. **134**(1), 127–133 (2018)
6. Chen, Y., Hui, B., Cheng, Y., Hao, Y., Fu, M.: Failure investigation of buffer layers in high-voltage XLPE cables. Eng. Fail. Anal. **113**, 104546 (2020). https://doi.org/10.1016/j.engfailanal.2020.104546
7. Abbasi, V.: Classifying faults locations in cable terminations and investigation of the faults reasons. Iran. J. Electr. Electron. Eng. **14**(3), 270–277 (2018). https://doi.org/10.22068/IJEEE.14.3.270
8. Surzhikov, A.P., Pritulov, A.M., Lysenko, E.N., Sokolovskiy, A.N., Vlasov, V.A., Vasendina, E.A.: Calorimetric investigation of radiation-thermal synthesized lithium pentaferrite. J. Therm. Anal. Calorim. **101**(1), 11–13 (2010)
9. Surzhikov, A.P., Peshev, V.V., Pritulov, A.M., Gyngazov, S.A.: Grain-boundary diffusion of oxygen in polycrystalline ferrites. Russ. Phys. J. **42**(5), 490–495 (1999)
10. Surzhikov, A.P., Frangulyan, T.S., Ghyngazov, S.A., Lisenko, E.N., Galtseva, O.V.: Physics of magnetic phenomena: investigation of electroconductivity of lithium pentaferrite. Russ. Phys. J. **49**(5), 506–510 (2006)

11. Andrade, A.F., Costa, E.G., Andrade, L.M., Soares, S.H., Lira, R.S.: Design of cable termination for AC breakdown voltage tests. Energies **12**(16), 3075 (2019). https://doi.org/10.3390/en12163075
12. Starikova, N.S., Redko, V.V., Vavilova, G.V.: Control of cable insulation quality by changing of electrical capacitance per unit during high voltage testing. J. Phys.: Conf. Ser. **671**(1), 012056 (2015). https://doi.org/10.1088/1742-6596/671/1/012056
13. Goldshtein, A.E., Vavilova, G.V., Belyankov, V.Y.: An electro-capacitive measuring transducer for the process inspection of the cable capacitance per unit length in the process of production. Russ. J. Nondestr. Test. **51**(2), 86–93 (2015). https://doi.org/10.1134/S1061830915020047
14. Zou, X., Mu, H., L. Qu, Zhang, H., Xie, C., Zhang, G.: Localization and assessment of breakage defect in cables based on time–frequency domain reflectometry. Energy Rep. **8**(5), 1474–1481 (2022). https://doi.org/10.1016/j.egyr.2022.02.202
15. Alonso, G., Meseguer, J., Sanz-Andres, A., Valero, E.: On the galloping instability of two-dimensional bodies having elliptical cross-sections. J. Wind Eng. Ind. Aerodyn. **98**, 438–448 (2010). https://doi.org/10.1016/j.jweia.2010.02.002
16. Afia, A., Ehtasham, M., Zoltan, T.: Dielectric spectroscopy of low voltage nuclear power cables under simultaneous thermal and mechanical stresses. Energy Rep. **6**, 662–667 (2020). https://doi.org/10.1016/j.egyr.2020.11.155
17. Chursin, Y.A., Redko, L.A., Fedorov, E.M: Enlargement of measuring zone in laser gauges without sacrificing measurement accuracy. Meas.: J. Int. Meas. Confed. **131**, 647–653 (2019). https://doi.org/10.1016/j.measurement.2018.09.031
18. Feng, B., Zhang, L., Hou, S., et al.: Research on cable defect location method based on joint time-frequency analysis. In: IEEE International Conference on Electrical Materials and Power Equipment, pp. 1–4 (2021)
19. Eigner, A., Semino, S.: 50 years of electrical-stress control in cable accessories. IEEE Electr. Insul. Mag. **29**(5), 47–55 (2013). https://doi.org/10.1109/MEI.2013.6585856
20. Huang, F.: Cause analysis and countermeasures of a breakdown failure of flexible 110 kV cable terminal. In: E3S Web of Conferences, vol. 38, pp. 2–6 (2018). https://doi.org/10.1051/e3sconf/20183804004
21. Fedorov, E.M., Koba, A.A.: Three-axis laser method for measuring the diameter of cylindrical objects. Dyn. Syst. Mech. Mach. 7819008 (2016). https://doi.org/10.1109/Dynamics.2016.7819008
22. Chimunda, S., Nyamupangedengu, C.: A reliability assessment model for an outdoor 88 kV XLPE cable termination. Electr. Power Syst. Res. **177**, 105979 (2019). https://doi.org/10.1016/j.epsr.2019.105979
23. Mazzanti, G., Marzinotto, M.: Advanced electro-thermal life and reliability model for high voltage cable systems including accessories. IEEE Electr. Insul. Mag. **33**(3), 17–25 (2017). https://doi.org/10.1109/MEI.2017.7906159
24. Fynes-Clinton, D., Nyamupangedengu, C.: Partial discharge characterization of cross-linked polyethylene medium voltage power cable termination defects at very low frequency (0.1 Hz) and power frequency test voltages. IEEE Electr. Insul. Mag. **32**(4), 15–23 (2016). https://doi.org/10.1109/MEI.2016.7528986
25. Linde, E., Verardi, L., Pourmand, P.: Non-destructive condition monitoring of aged ethylene-propylene copolymer cable insulation samples using dielectric spectroscopy and NMR spectroscopy. Polym. Test. **46**, 72–78 (2015). https://doi.org/10.1016/j.polymertesting.2015.07.002
26. Zhang, H., Mu, H., Lu, X., Tian, J., Zou, X.: A method for locating and diagnosing cable abrasion based on broadband impedance spectroscopy. Energy Rep. **8**, 1492–1499 (2022). https://doi.org/10.1016/j.egyr.2022.03.163

Technology of Gold Extraction from Clay Ore and Technogenic Raw Materials

Sergey Bordunov, Yuriy Fedorchuk, Olga Galtseva, and Gufana Narimanova

Abstract Currently technogenic gold-bearing raw materials (tailings and dumps of existing or closed mines) are considered by many experts as a promising raw material for the extraction of residual gold. Technogenic raw materials, as a rule, are crushed, have fairly clear boundaries and are located near energy sources, roads and settlements, which minimize capital costs for its processing. A negative factor that hinders the recycling of technogenic raw materials is the presence of substances harmful to human health in this raw material: floating and leaching reagents, mercury, etc. If the content of clay and sludge fractions in the raw material is 20 –30%, there is a structural bonding of particles and sorption of finely dispersed gold on clays, which makes it difficult to extract. The extraction of finely dispersed gold from hard-to-wash clay rocks and technogenic raw materials is an important task.

S. Bordunov
Public Corporation Scientific - Innovation Enterprise "Echteh", 4/3 Akademichesky Ave., Tomsk 634055, Russia

Y. Fedorchuk
Division for General Technical Subjects, National Research Tomsk Polytechnic University, 30 Lenin Avenue, Tomsk 634050, Russia
e-mail: ufed@mail.ru

O. Galtseva (✉)
School of Non-Destructive Testing, National Research Tomsk Polytechnic University, 30 Lenin Avenue, Tomsk 634050, Russia
e-mail: piano@tpu.ru

G. Narimanova
Department of Innovation Management, Tomsk State University of Control Systems and Radioelectronics, 40, Lenin Avenu, Tomsk, Russia 634050
e-mail: guftana@mail.ru

E. Lysenko et al. (eds.), *Emerging Trends in Materials Research and Manufacturing Processes*, Engineering Materials, https://doi.org/10.1007/978-3-031-38964-1_4

1 Introduction

Clays are able to form dense, "shell-like" and impervious to water "coats" (including particles of free gold) on detrital grains [1, 2]. Such gold is not available for aqueous leaching solutions. In addition, the "coat" sharply reduces the density of the gold particle, enclosed in it, and the hit probability of such a particle into the gravitational concentrate. In practice, raw materials with the clay content more than 3% are clay rocks, and raw materials with this contain more than 30% are clays. Almost all clays are polymineral rocks, consisting of a mixture of hydrosilicate minerals and impurities of non-clay minerals: quartz, feldspar, oxidized iron minerals, etc.

The works [3–6] proposed electroexplosive processing of clay raw materials with energy consumption of 4–10 kJ/kg for disintegration of clays to a finely dispersed state, at which the clay particle size becomes less than 6 μm; the deposition rate of such particles in the laminar mode becomes less than the deposition rate of gold particles with a size of 1 micron. The energy of such a discharge is comparable to the explosive explosion (EE), depending on the EE type with mass of 50 –200 g.

Works [7–10] also show the need for disintegration and removal of clay components from raw materials. Thin gold can be extracted only after the removal of clay from such placers. It is quite obvious that the possibility of extracting fine gold will appear only if there is a technology for the destruction of clay agglomerates to a particle fineness of less than 6–10 μm, both from ore and technogenic raw materials.

According to [11–18], this problem can be solved by electroexplosive dispersion of the clay component of raw materials in aqueous suspensions to ultra- and microaggregate associations of particles.

2 Theoretical Part

The energy of the discharge is used in the process of electroexplosive enrichment of mineral raw materials. Electric discharge in a liquid (EDL) is characterized by high energy concentration, high pressure and temperature. The liquid breakdown time is from 0.1 to 2 μs and depends on the operating voltage of the discharge circuit. The duration of the electric discharge is determined by the capacitance of the capacitor bank and the inductance of the discharge circuit. The duration of the electric discharge is determined by the capacitance of the capacitor battery and the inductance of the discharge circuit. The channel is filled with liquid plasma with a density of up to 10^{20} particles per 1 cm^3, which ensures its heating to 4 –104 K and the plasma existence time of at least 10 –4 s. A vapor–gas cavity is formed with a pressure of up to 10^2 -10^3 MPa after the breakdown of the liquid. It accounts for up to 31% of the energy released in the discharge channel.

A shock wave is formed, which is a significant destructive force at the instantaneous expansion of the discharge channel. The pressure at the shock wave front reaches 10^9 Pa. Approximate energy distributions during EDL playback are as follows: shock wave energy is 63%, gas–vapor cavity is 31% and light radiation is 6%. An electric discharge in a liquid is accompanied by complex physicochemical phenomena with the formation, for example, in water, of such ionization products as H^+, OH^-, O^*, O_3. In total, eleven active ions were recorded in the products of water ionization, which are able to oxidize sulfide sulfur to S^{6+}.

Compression shock waves, a pulsating vapor–gas cavity, electromagnetic, light and UV radiation and secondary cavitation in a rarefaction wave are formed as a result of the discharge. Thus, an electrical discharge in a liquid is a means of directly converting most of the electrical energy into mechanical energy.

A discharge occurs with the generation of a shock wave, cavitation phenomena and hydraulic flows that affect the material being processed when a high-voltage pulsed voltage is applied to the electrodes. Wherein, a net of microcracks is formed in the rock material first (mainly along the interfaces between individual phases and inclusions of minerals); the hard rock sample is destroyed when deformations accumulate above the limit strength. Multiple cavitation contributes to the crushing of the material and the violation of its structural order during the "collapse" of air bubbles [19, 20].

Figure 1 shows micrographs taken on an MBS-9 microscope of the initial clay raw material and the resulting gold concentrate without clay inclusions.

First, the results of the complete removal of clays from the gold-bearing raw materials of the Olkhovskoye deposit were published in our research [12]. The method of electroexplosive processing using electric discharges with an energy of 45–120 J was used to disperse clays with their subsequent removal from the process and extract fine gold from clay raw materials. The average gold grade in the raw material was 2.08 g/t; moreover, more than 40% of gold was represented by particles with a size of less than 0.05 mm.

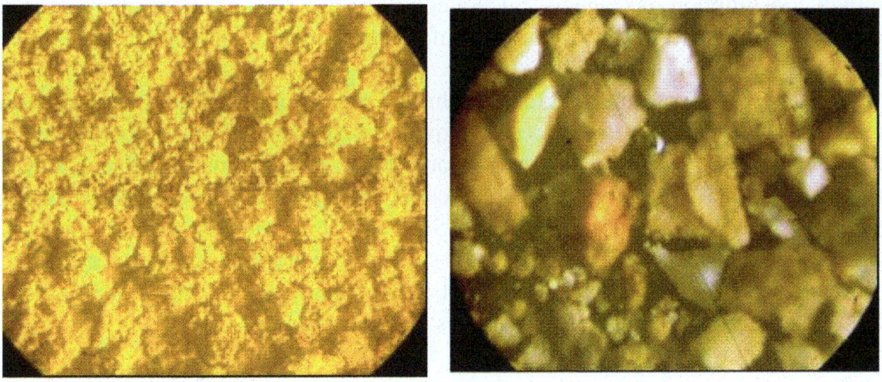

a) b)

Fig. 1 Microphotos of raw materials from the Olkhovskoye deposit and **a** the concentrate obtained from it after treatment with 20 discharges of 40 J, where a is feedstock; **b** is concentrate without clay inclusions

Gold concentrates with a mass of 9–20% by weight of the feedstock were obtained after electroexplosive enrichment of 2200 kg of raw materials; the degree of gold extraction from them was 97%. Such a high degree of gold recovery was obtained due to the almost complete recovery of fine gold with a size of particles of less than 0.05 mm, which are free from clay "coats".

The main advantages of the electroexplosive technology for the enrichment of gold-bearing raw materials are the almost complete removal of clays from the raw materials without loss of gold with a particle fineness of more than 10 μm and the availability of tested electroexplosive equipment with a productivity up to 100 m^3/h.

3 Experimental Part

In this work, the clayey tailings of gravity enrichment of the Berikulskaya Kemerovo region were choosen as the object of study by us, hereinafter referred to as "B" sample. The Berikulskaya mine is one of the oldest mines in Russia, formed as early as 1811. The mass of tailings is probably several million tons, of which about 400,000 tons of cyanidation cakes are compactly stored—waste from the last years of the mine's operation. From 1974 to 1983 attempts of "Irgiredmet" and the Laboratory of the Berikul gold extraction factory (GEF) have been repeatedly made to extract gold from these cakes; the weighted average content of gold and silver in them were 4.9 and 20.7 g/t, respectively. The main mass of gold is represented by particles with a fineness of 10 –50 μm. The best result of them is the degree of gold recovery of 60 –62% at the consumption of NaCN and $Ca(OH)_2$ on 15 –17 kg and 70 –80 kg per 1 ton of cakes, respectively. The process was deemed unprofitable. It is noteworthy that the content of clays in cakes is 40–50%; it is their presence that has caused such a high consumption of reagents. Such a complex object aroused our interest for an additional assessment of the possibility of electroexplosive enrichment of clay raw materials containing fine gold.

The paper [13] presents diagrams of three types of industrial and pilot industrial electroexplosive reactors for processing aqueous suspensions from fine particles with a particle fineness of less than 0.63 to coarse particles with a particle fineness of up to 10 cm. The productivity of such reactors is up to 100 m^3 by aqueous suspension with a volume ratio for "solid: liquid" (S:L) = 1:5; the consumption of electrical energy does not exceed 1.5 kW · h/t of solid raw materials. A laboratory reactor with one electrode cell and a load of a solid raw material sample up to 300 g is used to determine the optimal operating modes of these reactors.

The main tasks of laboratory research are:

- determination of the amount of gold with a particle size of ≤10 μm, carried into tailings to assess the performance of an industrial plant in terms of the "solid: liquid" ratio in suspension and, ultimately, water consumption;
- determination of the mass of gold particles, "shielded" inside rock grains.

Water and air with flow control are introduced into the reactor tangentially from below. An aqueous suspension with a finely dispersed solid phase, the product of grinding larger grains of raw materials, is discharged from the upper branch pipe into a conical sump with a capacity of 60 dm^3. A material balance is compiled according to the mass of tailings in the sump and the mass of the remaining concentrate in the reactor. The fine gold particle fineness is determined; one of the types of pilot reactors is selected for processing an experimental batch of raw materials based on the water flow rate through the reactor and the gold content in the tailings of the sedimentation tank.

Preliminary technological studies on the electrohydraulic treatment of "B" sample on a laboratory stand were carried out in detail with a water supply rate to the reactor equal to the rate of deposition of a gold particle with a fineness of 10 μm. As a result, we were able to disperse and isolate up to 40% of clay components into tailings with running water. The loss of gold with tailings with a particle fineness of less than 10 μm did not exceed 9.5% of the mass of gold in the feedstock with a gold grade of 1.81 g/t. The energy consumption of electroexplosive enrichment associated with the dispersion of clay components and their removal from the process amounted to 1.5 kW · h/t of cakes. The results of electroexplosive enrichment of "B" sample are presented in Tables 1 and 2.

The water flow regime with flow rate of 3.6 m/h can be considered practically laminar, neglecting turbulence in short-term discharge time periods (10^{-6} –10^{-12} s) with a frequency of 1 Hz. Nevertheless, some particles with fineness more than 45 μm ejected by shock waves into the outlet pipe during these periods, although quartz particles with a fineness of no more than 40 μm can be carried out simultaneously with gold particles with a fineness of 10 μm at a water flow rate of 3.6 m/h in a laminar stream.

The mass loss of the sample after electroexplosive treatment is associated with not only the removal of fine particles with running water and with the presence of soluble salts and surfactants in the cakes. As a foam is in the flowing water at the outlet of the reactor, then the electrical conductivity of the water increases by 2 –3 times and the pH decreases from 7.1 to 6.4. The share of large particles in the tailings

Table 1 Granulometric composition of the concentrate and tailings of "B" sample after electroexplosive enrichment

Fraction, (mm)	Concentrate, (%)	Tailings, (%)	Losses, (%)
0.315 < 2.5	18.77	–	–
0.315 < 0.45	8.57	0.02	–
0.16 < 0.315	16.91	0.08	–
0.1 < 0.16	8.24	0.44	–
0.071 < 0.1	5.09	2.95	–
0.063 < 0.071	1.67	2.7	–
0.045 < 0.063	0.5	7.43	–
<0.045	–	19.95	6.68
–	59.75	33.57	–

Table 2 Phase composition of "B" sample, concentrate and tailings and gold balance after electroexplosive treatment in a laboratory reactor

Mineral	Content by volume, (%)		
	Initial	Concentrate	Tailings
Quartz SiO_2	35	60	61
$Ca(SO_4)(H_2O)_{0,5}$	9	–	–
$KMg_3(Si_3Al)O_{10}F_2$	17	5	15
$(K, H_2O)Fe_3(SO_4)_2(OH)_6$	12		5
FeS	19	5	–
TiO_2, FeS, Kinochlor-$(Mg, Fe)_6(Si, Al)_4O_{10}(OH)_8$,	<5	–	–
$Na(AlSi_3O_8)$, $Ca(SO_4)(H_2O)_{0,5}$, $(K, H_2O)Fe_3(SO_4)$	–	<5	–
Kinochlor	–	6	2
$Ca(SO_4)$	–	7	–
TiO_2	–	8	–
$Fe_3(OH)(SO_4)2H_2O$, $(Na, Ca)Al(Si, Al)_3O_8$, FeS, $Ca(SO_4)(H_2O)_{0,5}$	–	–	<5
TiO_2, FeS, $Fe_3(OH)(SO_4)2H_2O$, $Ca(SO_4)(H_2O)_{0,5}$, $Ca(SO_4)$	–	–	7
Sum, % of 100%	97	96	95

is also a probabilistic estimate of the error in calculating the technological yield of gold by this method. Since the settling rate of a quartz particle with a fineness of 0.45 mm corresponds to the settling rate of a gold particle with a fineness of ~0.0115 mm, the probability of ejection of such a particle into tailings is 0.02%. However, the proportion of gold with a particle fineness of less than 10 μm that ended up in tailings is only 4.667%. The bulk of the gold (94.21%) remained in the concentrate.

An experiment to determine the presence of gold particles "shielded" by the rock in samples was performed on 100 g samples by direct leaching of gold with "aqua regia", twice repeated washing of the solid precipitate with distilled water with a volume of 50 ml, combining the liquid phase and mass spectral analysis of the liquid phase for gold.

The washed precipitate after first leaching was treated in an electroexplosive reactor with 30 discharges with an energy of 112.5 J and gold was repeatedly leached from them as described above. The gold grade was 1.57 g/t, after the second leaching it was 0.34 g/t and the total gold grade was 1.91 g/t after the first leaching. Thus, in "B" sample 17.8% of gold turned out to be enclosed within the grains of the accompanying rock and inaccessible to "aqua regia" without destroying these grains without electroexplosive treatment. It is possible that this is also the reason for the low yield of gold during direct leaching of cakes with sodium cyanide with lime, as noted above.

The turbulent mode of obtaining gravitational concentrates, in which the settling rates of equal-sized particles of gold, pyrite and quartz are related as 5:2:1, is one of the main reasons for the low extraction of gold with a particle fineness of less than 50 μm by the gravitational method in modern apparatuses. The deposition rates of gold, pyrite and quartz are related as 14:3:1 in the laminar mode. Based on this, the fine gold concentrate itself was obtained in the laminar mode of the suspension flow in the deposition chamber connected to the reactor after the electroexplosive processing of raw materials.

Figure 2 shows a diagram of a reactor with the supply of solid raw materials with natural moisture from top to bottom, towards the upward flow of water, into which, if necessary, air can also be introduced through the ejector. Heavy particles, the settling speed of which is higher than the speed of the upward flow of water, fall into the collection hopper, from which the resulting heavy concentrate is removed to the outside by an elevator. Particles that do not settle in the upward flow of water (tails) are removed from the reactor through nozzles installed in the upper part of the reactor. Grounded electrodes are shelves installed inside the reactor at an angle of 30 –450 to the vertical axis of the reactor. Due to the sliding of the solid phase along the shelves, the time of its stay in the reactor increases, due to the vibration of the shelves under the action of shock waves and springs, even the smallest solid particles slide along them. The design of the shelves in the upper part of the reactor ensures the exit of a fine suspension of tailings without pulsations and in a regime close to laminar into a special chamber for trapping fine gold with a particle fineness of 50 μm or less. The experiment in this reactor was carried out at a water upflow speed of 1.45 m/min. At this speed, gold particles with a particle size of 50 μm or less were carried out of the reactor into the deposition chamber together with quartz particles with a size of less than 190 μm.

The second mode of the experiment was carried out at an upward water flow rate of 5.69 10^{-2} m/min, at which a finely dispersed suspension of clays and gold with a particle size of less than 10 μm is carried into the chamber, i.e. during the operation of the laboratory experiment. The limit for trapping gold particles with a

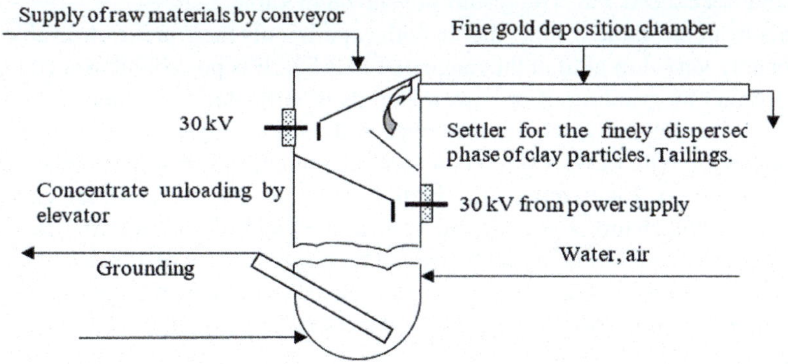

Fig. 2 Pilot electroexplosive plant for extracting fine gold into concentrate from clay raw materials

size of 10 μm is set based on the rate of deposition of gold particles. Thus, gold particles with fineness of 1 μm are deposited 100 times slower than 10 μm particles (proportional to the square of the diameter), and this requires increasing the length of the deposition chamber by 100 times. It is not expedient and not profitable due to the low content of such gold in cakes. The product obtained in the chamber can be further enriched by classification on sieves to remove a large fraction of the accompanying rock from it.

Two dry samples "B" of 50 kg each were selected for the experiment on a pilot plant (Fig. 2). The first sample (experiment no. 1) was processed at a frequency of 2 Hz (~36 discharges with an energy of 112.5 J for each particle falling into the receiving hopper) at an upward water flow rate of 5.69 10^{-2} m/min.

It corresponded to no entrainment in tails of gold particles of 10 or more microns and the feed rate of cakes to the reactor by the conveyor is 5 dm^3/min, which corresponded to the ratio T:L in the upper part of the reactor ≈1:5. This consumption of raw materials corresponded to the plant productivity of 0.79 t/h. The tailings were passed through a cone-shaped sump with an apex angle of 60° from the laboratory stand, into which the suspension was introduced tangentially from below, and partially clarified water was discharged tangentially from the upper, wide part of the sump into the sewer. The precipitate was analyzed.

The second experiment (no. 2) was performed at a frequency of 2 Hz at an upward water flow rate of 1.45 m/min, slightly higher than the calculated sedimentation rate of 50 μm gold particles, equal to 1.42 m/min. The feedstock consumption was set equal to 10–13 dm^3/min, or 1.58–2.1 t/h to maintain the ratio T:L ≈ 1:5 in the upper part of the reactor. A gold precipitation chamber was attached to the reactor as shown in Fig. 2. The water supply to the reactor was turned off at the end of the experiment, the heavy concentrate was removed from the lower hopper and the mat with sediment from the gold precipitation chamber. The concentrates from the hopper, sediment on the mat and tailings from the settling tank were dried and analyzed for gold by mass spectral and neutron activation methods of analysis. The result was the average of three analyses. The convergence of the results for both methods of analysis was quite satisfactory. Deviations of the results of the analysis from the average value of the three did not exceed 7%. The results are shown in Table 3.

Data from the Table 3 show that in both experiments there are irretrievable losses of probably very thin gold in the range of 1–1.65%. It is possible that such particles were formed by grinding of ore raw materials in mills due to the abrasion of larger particles. The yield of gold into the concentrate of experiment no.1 is 89.62% with a plant productivity of 0.79 t/h. 93.44% of the initial gold was collected on the mat at experiment no. 2 with a plant productivity of 1.58–2.1 t/h. Thus, the expediency of installing the chamber is undoubted, since the yield of gold into the concentrate increases and the productivity of the reactor increases by 2–2.66 times. Sampling for analysis of gold from this concentrate was carried out after screening out a fraction larger than 1.1 mm, the mass of which was 6.35 kg or 12.7% of the mass of the feedstock.

Table 3 The results of electroexplosive processing of two "B" samples of tailings at a water flow rate of 5.69 10^{-2} m/min without a gold deposition chamber (no.1) and at a water flow rate of 1.45 m/min with a gold deposition chamber (no. 2)

	Material balance, (kg) and (%)			
	Mass of concentrate, (kg/%)	Mass of sediment on the mat, (kg/%)	Mass of tailings in the sump, (kg/%)	Loss of mass in the experiment, (kg/%)
1	27.34/54.68	–	8.12/16.24	14.54/29.08
2	20.05/40.1	19.94/39.88	–	10.01/20.02
Gold balance, (g/t) and (%)				
1	1.64/89.62	–	0.17 /9.29	0.02/1.09
2	0.09/4.91	1.71/93.44	–	0.03/1.65

4 Conclusion

The paper shows the fundamental possibility of cost-effective processing of difficult-to-wash clayey gold-bearing ores using a high-voltage discharge in aqueous suspensions with an energy consumption of not more than 1.5 kWh/t. The presented technological process was carried out using an electric explosive installation (Fig. 2) and made it possible to separate the initial sample into the following three products:

- tailings, represented by a clay fraction and fine fragments of other rocks, weighing 33.57% of the mass of the original sample with a gold content of 1.88% of the original;
- gold concentrate with a grain size of 10–50 μm (in terms of the volume of the ball), the mass of which was 25.4% by weight of the feedstock with a gold content of 96.22% of the original;
- coarse-grained fraction (41.35% of the original mass) with a gold content of 1.9% of the original gold in the sample.

The proposed method makes it possible to remove easily crushed rocks from clay raw materials with the production of small volumes of concentrate and, accordingly, makes it possible to drastically reduce the consumption of water and chemical reagents at subsequent stages of gold leaching. The method is also applicable to other types of gold-bearing raw materials.

References

1. Osipov, V.I., Sokolov, V.N., Rumyantseva, N.A.: Microstructure of Clay Rocks. Nedra, Moscow (1989)
2. Deschenes, G., Xia, C., Fulton, M., Cabri, L.J., Price, J.: Evaluation of leaching parameters for a refractory gold ore containing aurostibite and antimony minerals: Part I—Central zone. Miner. Eng. **22**(9–10), 799–808 (2009)

3. Zhizhaev, A.M.: Electroexplosive method of high-energy intensive processing of sulfide raw materials. Chem. Sustain. Dev. **6**(2–3), 8 (1998)
4. Bordunov, S.V., Galtseva, O.V.: About prospects of enrichment of mineral raw materials and chemical activation of water suspensions by electroexplosive method. J. Phys: Conf. Ser. **671**(1), 012009 (2016). https://doi.org/10.1088/1742-6596/671/1/012009
5. Lysenko, E., Nikolaev, E., Vlasov, V., Surzhikov, A.: Microstructure and reactivity of Fe_2O_3–Li_2CO_3–ZnO ferrite system ball-milled in a planetary mill. Thermochim. Acta **664**, 100–107 (2018)
6. Lysenko, E.N., Malyshev, A.V., Vlasov, V.A., Nikolaev, E.V., Surzhikov, A.P.: Microstructure and thermal analysis of lithium ferrite pre-milled in a high-energy ball mill. J. Therm. Anal. Calorim. **134**(1), 127–133 (2018)
7. Naumov, V.A.: Processes of formation and distribution of concentrations of noble metals in technogenic placers and dumps of the Urals. Min. J. Ural Mount. Rev. Ekaterinburg **8**, 39–50 (1994)
8. Surzhikov, A.P., Pritulov, A.M., Lysenko, E.N., Sokolovskiy, A.N., Vlasov, V.A., Vasendina, E.A.: Calorimetric investigation of radiation-thermal synthesized lithium pentaferrite. J. Therm. Anal. Calorim. **101**(1), 11–13 (2010)
9. Generalov, M.E., Naumov, V.A.: Transformation of gold in technogenic placers and dumps of the Urals. Ural. Geol. J. **4**, 19–56 (1998)
10. Surzhikov, A.P., Peshev, V.V., Pritulov, A.M., Gyngazov, S.A.: Grain-boundary diffusion of oxygen in polycrystalline ferrites. Russ. Phys. J. **42**(5), 490–495 (1999)
11. Gulyi, G.A.: Equipment and Technological Processes Using the Electrohydraulic Effect. Mashinostroenie, Moscow (1997)
12. Bordunov, S.V., Bordunov, V.V., Makarychev, Yu.I.: The technology of extracting gold from ore and technogenic raw materials. Russ. J. Non-Ferrous Metals **9**, 24–28 (2008)
13. Galtseva, O.V., Bordunov, S.V., Kulaga, I.G.: Electroexplosive equipment for dispersing, chemical activation and mineral processing. Tomsk Polytechnic University Publishing House, Tomsk (2013)
14. Bordunov, S.V., Makeev, V.A., Shepelev, I.T.: Electro-explosive technology for activation of lime milk. Russ. J. Non-Ferrous Metals **2**, 65–68 (2008)
15. Cabri, L.J., Beattie, M., Rudashevsky, N.S., Rudashevsky, V.N.: Process mineralogy of Au, Pd and Pt ores from the Skaergaard Intrusion, Greenland, using new technology. Miner. Eng. **18**(8), 887–897 (2005). https://doi.org/10.1016/j.mineng.2005.01.021
16. Cabri, L.J., Rudashevsky, N.S., Rudashevsky, V.N., Gorkovetz, V.Y.: Study of native gold from the Luopensulo deposit (Kostomuksha area, Karelia, Russia) using a combination of electric pulse disaggregation (EPD) and hydroseparation (HS) Original Research. Miner. Eng. **21**(6), 463–470 (2008). https://doi.org/10.1016/j.mineng.2008.02.006
17. Adams, M.D.: Advances in Gold Ore Processing. Elsevier, Netherlands (2005)
18. Goodall, W.R.: Characterisation of mineralogy and gold deportment for complex tailings deposits using QEMSCAN(R). Miner. Eng. **21**(6), 518–523 (2008). https://doi.org/10.1016/j.mineng.2008.02.022
19. Surzhikov, A.P., Frangylyan, T.S., Ghyngazov, S.A.: A dilatometric study of the effect of pressing on the kinetics of compression of ultrafine zirconium doxide powders under thermal annealing. Russ. Phys. J. **55**(4), 345–352 (2012). https://doi.org/10.1007/s11182-012-9818-1
20. Duan, C.L., Diao, Z.J., Zhao, Y.M., Huang, W.: Liberation of valuable materials in waste printed circuit boards by high-voltage electrical pulses. Miner. Eng. **70**, 170–177 (2015). https://doi.org/10.1016/j.mineng.2014.09.018

Magnetization Study of $Li_{0.5}Fe_{2.5}O_4$ Ferrite Synthesized by the Electron-Beam Heating of Mechanically Activated Fe_2O_3–Li_2CO_3 Mixture

Elena Lysenko, Evgeniy Nikolaev, Anatoliy Surzhikov, and Julia Minina

Abstract In this work, the magnetization of lithium ferrite synthesized from a mechanically activated Fe_2O_3–Li_2CO_3 initial reagents mixture via electron-beam heating was studied by thermomagnetometric analysis. Mechanical activation was carried out using the high-energy grinding of powders in an AGO-2S ball mill. The rotation speed of the steel vials and the grinding time were 1290, 2220 rpm and 5, 15, 30, 60, 120 min, respectively. The samples were synthesized using an ILU-6 pulsed electron accelerator with an electron energy of 2.4 meV. Under certain conditions, the electron beam allowed heating samples up to synthesis temperatures of 600–750 °C. For comparison, some of the samples were synthesized via conventional heating in a laboratory furnace under the same temperature and time conditions. Thermomagnetometric measurements based on thermogravimetry in an external magnetic field were performed at temperatures corresponding to magnetic phase transitions at the Curie point of ferrites. It has been established that the magnetization of the samples depends both on the energy intensity of grinding and on the method of heating. The results revealed a higher degree of magnetization of the samples pre-activated at 2220 rpm for 120 min and then synthesized in an electron beam.

E. Lysenko (✉) · E. Nikolaev · A. Surzhikov · J. Minina
National Research Tomsk Polytechnic University, 30 Lenin Avenue, Tomsk, Russia 634050
e-mail: lysenkoen@tpu.ru

E. Nikolaev
e-mail: nikolaev0712@tpu.ru

A. Surzhikov
e-mail: surzhikov@tpu.ru

J. Minina
e-mail: ysm7@tpu.ru

E. Lysenko et al. (eds.), *Emerging Trends in Materials Research and Manufacturing Processes*, Engineering Materials, https://doi.org/10.1007/978-3-031-38964-1_5

1 Introduction

Magnetically soft lithium ferrites with $Li_{0.5}Fe_{2.5}O_4$ chemical composition have a spinel structure and are characterized by high values of saturation magnetization 60–68 emu/g and Curie temperature 630 °C, as well as low coercive force 20–85 Oe [1–3]. Such ferrites are widely used in radio electronics and instrument engineering [4–6].

It is known that the magnetic properties of spinel compounds are due to the exchange interaction between the spin magnetic moments of neighboring cations with unoccupied 3d-levels. Therefore, according to the Neel theory [7], the spontaneous magnetization of spinel ferrites is realized due to indirect exchange interaction between cations located in different sublattices (with the participation of oxygen ions). In this case, the resulting magnetic moment is determined by the difference between the magnetic moments of the octahedral and tetrahedral sublattices. It is known [8] that lithium ferrites have high values of spontaneous magnetization, which is formed at the stage of ferrite synthesis.

Currently, there are several ways to obtain ferrites, including solid-phase synthesis, sol-gel technology, microwave synthesis, etc. [9–12]. In this case, the main task of the above methods is to obtain ferrites with specified, reproducible and uniform electromagnetic properties.

The traditional solid-phase synthesis of ferrites is based on thermal heating of compacted samples in furnaces [13, 14]. In the case of obtaining lithium ferrite [15], during the synthesis, solid-phase interaction occurs between the initial iron oxide, Fe_2O_3, and lithium carbonate, Li_2CO_3, at high temperatures. The rate and temperature of such interaction depends on the activity of the powder mixture, which can be increased as a result of its mechanical grinding in a ball mill [16–19].

It was shown in [20–23] that the rate of solid-phase interaction between initial reagents of ferrite can also be increased by using an electron beam as a source of sample heating during their synthesis. In this case, the electron energy in the range of 1–3 meV makes it possible to heat compacted powders to a depth of 2–4 mm, depending on the density of the samples [24].

For Li ferrite synthesis under conditions of electron-beam heating, the main interaction between Fe_2O_3–Li_2CO_3 occurs at the stage of sample heating to the isothermal holding temperature [25]. From XRD data, the analysis of spinel phase formation during the ferrites synthesis showed that the accelerated formation of ferrite under conditions of electron-beam heating is associated with a decrease in the activation energy of synthesis and a decrease in the pre-exponential factor of the temperature dependence of the reaction rate.

It is known [26] that during the $Li_{0.5}Fe_{2.5}O_4$ lithium ferrite synthesis, a $LiFeO_2$ transient non-magnetic phase can form, which is difficult to separate from the ferrite phase in X-ray diffraction patterns. In this regard, in this work, the magnetization of lithium ferrite synthesized by the electron-beam heating of a mechanically activated Fe_2O_3–Li_2CO_3 initial reagents mixture was studied.

2 Research Method

To obtain $Li_{0.5}Fe_{2.5}O_4$ lithium ferrite, the initial pure Fe_2O_3 (91.5 wt%) and Li_2CO_3 (8.5 wt%) powders were preliminarily dried and mixed in an agate mortar. Then the powder mixture of reagents was subjected to mechanical grinding in an AGO-2S ball mill using steel vials and balls (6 mm in diameter) and vial rotation speeds of 1290 (MA1 samples) and 2220 (MA2 samples) rpm. The grinding time was 5, 15, 30, 60 and 120 min. The mass ratio of powder and balls was 1:10.

The mixture of initial reagents was divided into two batches. Samples from the first batch were synthesized by thermal heating in a laboratory furnace at temperatures of 600 and 750 °C for up to 120 min (T-samples). Samples from the second batch were synthesized under conditions of electron-beam heating (RT-samples) using the ILU-6 accelerator, which was developed at the Institute of Nuclear Physics SB RAS (Novosibirsk, Russia) [27–29]. This accelerator with 2.4 meV electron energy allow heating the samples being studied to synthesis temperature without any additional heat sources. In this case, the pulsed beam current, pulse duration, and pulse repetition rate were 400 mA, 500 μs, and 12.5 Hz, respectively.

The magnetization of the synthesized ferrite was analyzed by the results of thermomagnetometric measurements using a Netzsch STA 449C Jupiter thermal analyzer and the Netzsch Proteus Analysis software product. It was shown in [30] that thermomagnetometric measurements based on thermogravimetry in a magnetic field allow to analyzed magnetic phase transitions in ferrite samples and to determine their Curie temperature. Using calibration curves for control samples, it is possible to estimate the concentration of the magnetic phase in the test sample from the analysis of the magnetic phase transition "magnetic-paramagnetic" [30].

3 Results and Considerations

The synthesis of lithium ferrite occurs in accordance with the equation:

$$Li_2CO_3 + 5Fe_2O_3 \rightarrow 4Li_{0.5}Fe_{2.5}O_4 + CO_2 \tag{1}$$

However, during solid-phase interaction between the Fe_2O_3–Li_2CO_3 initial reagents up to 620 °C, both the lithium $Li_{0.5}Fe_{2.5}O_4$ ferrite and the non-magnetic phase $LiFeO_2$ are formed in accordance with parallel reactions:

$$Li_2CO_3 + Fe_2O_3 \rightarrow 2LiFeO_2 + CO_2 \tag{2}$$

$$Li_2CO_3 + 5Fe_2O_3 \rightarrow 4Li_{0.5}Fe_{2.5}O_4 + CO_2 \tag{3}$$

At a temperature of 690 °C and above, the process of $LiFeO_2$ interaction with Fe_2O_3 is activated in accordance with the reaction:

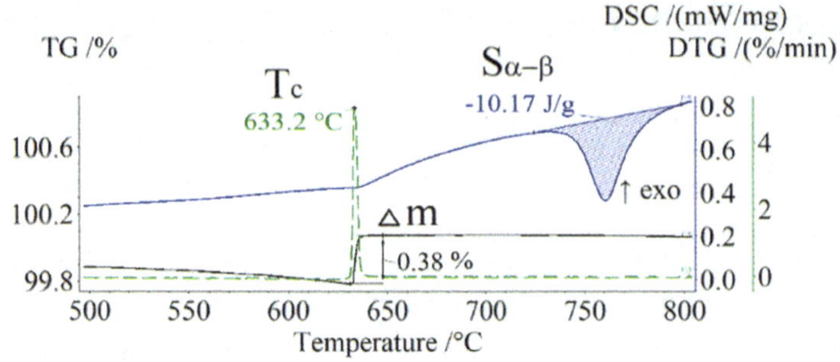

Fig. 1 Thermogravimetric (in a magnetic field) and differential scanning calorimetric analyzes for the control sample of lithium ferrite synthesized at 900 °C

$$LiFeO_2 + 2Fe_2O_3 = 2Li_{0.5}Fe_{2.5}O_4 \quad (4)$$

Thus, the rate of synthesis of $Li_{0.5}Fe_{2.5}O_4$ lithium ferrite sharply increases at high temperatures, while the amount of $LiFeO_2$ decreases.

Figure 1 shows thermogravimetric (TG) curves obtained by applying an external magnetic field, as well as differential scanning calorimetric (DSC) curves for a control sample synthesized by traditional thermal heating at 900 °C for 4 h. This analysis was carried out up to a temperature of 800 °C at a fast rate (50 K/min) to eliminate possible chemical transformations in the sample that affect the weight changes during thermomagnetometry.

As follows from Fig. 1 that when the synthesized sample is heated in a magnetic field, a weight jump (Δm) is recorded in the region of the ferrimagnet-paramagnet phase transition temperature, which is associated with the interaction of the ferrite with an external magnetic field. The magnitude of the weight change depends on the concentration of the synthesized magnetic phase, that is, the higher the content of the ferrite phase in the sample, the greater the weight jump. The resulting value of 0.38% can be taken as a calibration value for a sample with a ferrite content of 100 wt%. Using a derivative thermogravimetric curve (DTG), one can estimate the temperature of a given transition, that is, the Curie temperature of the ferrite. The value of 633.2 °C is close to the literature data for lithium ferrite with $Li_{0.5}Fe_{2.5}O_4$ composition [25].

In the area of the above-described magnetic phase transition, the DSC curve shows a small diffuse endothermic peak associated with the disordering of the magnetic structure of the ferrite, accompanied by a thermal effect. At a higher temperature, equal to ~760 °C, the DSC curve shows a peak with an area of 10.17 J/g, which reflects the transition of the lithium ferrite ordered phase (α-$Li_{0.5}Fe_{2.5}O_4$) to the disordered phase (β-$Li_{0.5}Fe_{2.5}O_4$) [25]. The area of this peak can also be taken as a calibration value for a sample with 100 wt% ferrite content.

Figures 2 and 3 show TG and DSC curves obtained for samples synthesized at 600 and 750 °C for 120 min from powders mechanically activated at 1290 rpm (Fig. 2) and 2220 rpm (Fig. 3). It can be seen that the height of the weight jump during the magnetic phase transition and the DSC peak area ($S_{\alpha \rightarrow \beta}$) of the $\alpha \rightarrow \beta$ transition in the $Li_{0.5}Fe_{2.5}O_4$ phase, which characterize the concentration content of lithium ferrite, depend on the synthesis temperature, the method of heating, and the activity of the samples. A variation in the Curie temperature is also observed. A more detailed change in the above parameters for all the samples is presented in Figs. 4 and 5, which summarize the data for all synthesis modes and type of samples.

The results indicate the formation of lithium ferrite with a significantly higher concentration from mechanically activated powders compared to the synthesis of ferrite from non-mechanically activated reagents at 600 °C (Fig. 4a), both under thermal and electron-beam heating conditions. However, with an increase in the synthesis temperature to 750 °C (Fig. 4b), the formation of lithium ferrite in the isothermal stage weakly depends on the method of heating and the degree of activity

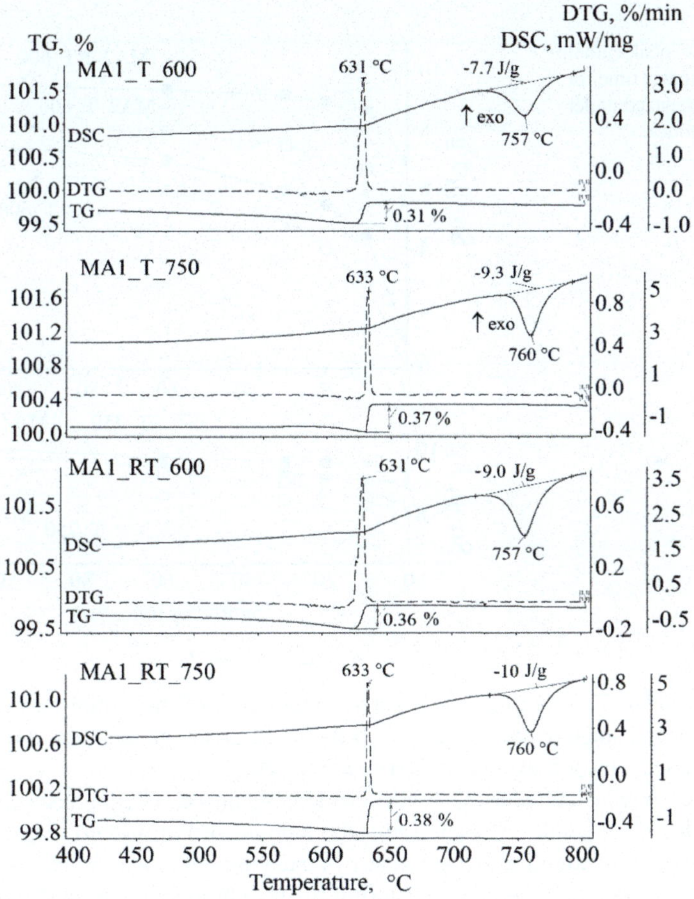

Fig. 2 Thermogravimetric (in a magnetic field) and differential scanning calorimetric analyzes of lithium ferrite synthesized in various modes from powders mechanically activated at 1290 rpm

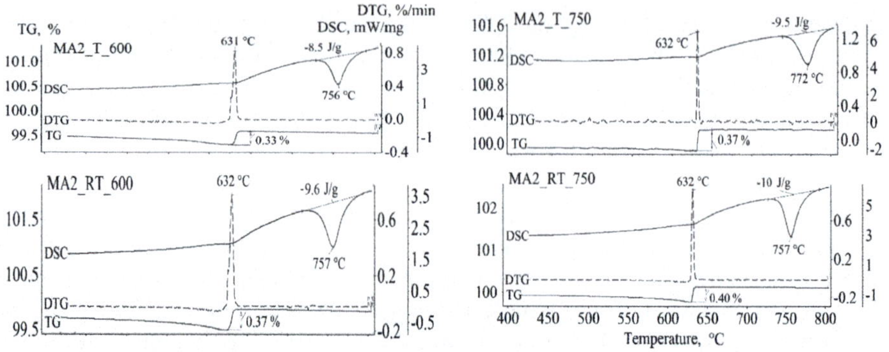

Fig. 3 Thermogravimetric (in a magnetic field) and differential scanning calorimetric analyzes of lithium ferrite synthesized in various modes from powders mechanically activated at 2220 rpm

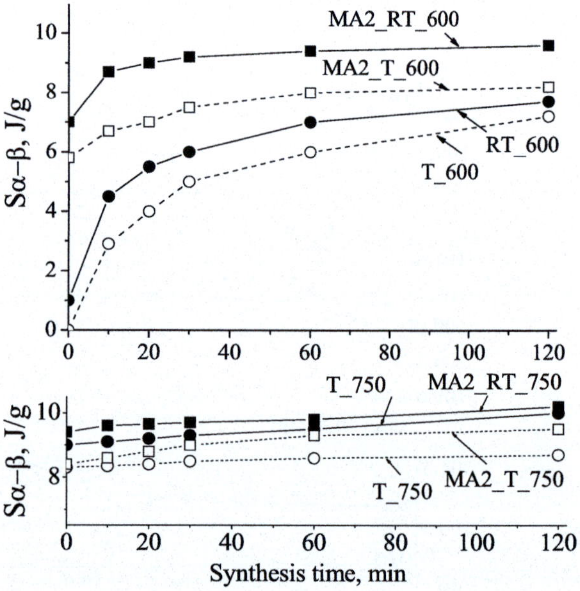

Fig. 4 DSC peak enthalpy versus synthesis time for ferrite synthesized under different modes

of the powders. This is due to the active interaction of the initial reagents at the non-isothermal stage of synthesis, i.e., at the stage of sample heating, during which the main concentration of lithium ferrite is formed.

Kinetic analysis showed that with an increase in the RT synthesis time above 60 min, the $S_{\alpha\to\beta}$ values characterizing the amount of the formed magnetic phase vary slightly, indicating the inexpediency of carrying out longer annealing.

Figure 5 shows the results of measuring the weight jump (Δm) at thermomagnetometric analysis, the value of which characterizes the magnetization of the sample. For all samples, the Δm increases with increasing synthesis time, which indicates

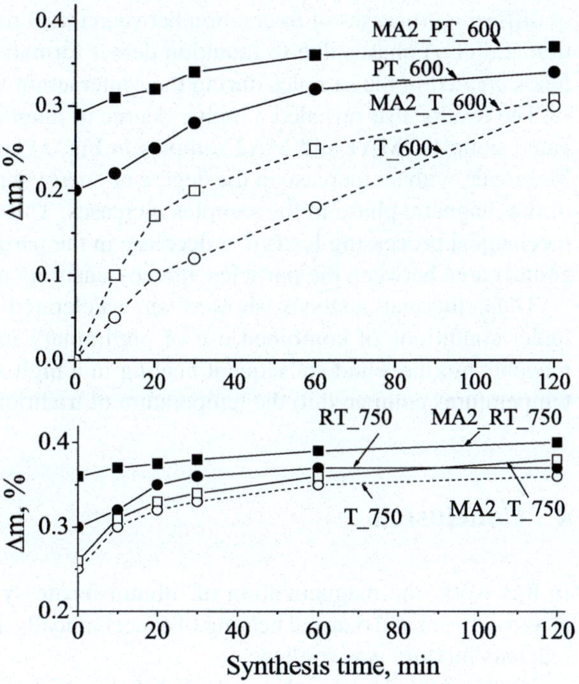

Fig. 5 The height of the weight jump versus synthesis time for ferrite synthesized under different modes

the accumulation of the magnetic phase in the samples. According to the XRD data obtained in [25], the behavior of the magnetization of the samples is similar to the change in the concentration of the spinel phase in such samples during their synthesis.

At a thermal synthesis temperature of 600 °C (T_600 samples in Fig. 5a), Δm of the samples increases strongly with an increase in the isothermal holding time, which is associated with the formation of lithium ferrite in accordance with Eqs. (2) and (4). However, in the case of sample synthesis using an electron beam (RT_600 samples in Fig. 5a), the Δm already has a high value for samples synthesized at short times. This indicates that the electron heating of the samples at 600 °C leads to a higher rate of diffusion interaction between the initial oxides with the formation of a magnetic ferrite phase during nonisothermal heating of the samples.

An increase in the synthesis temperature to 750 °C (T_750 and RT_750 samples in Fig. 5b) accelerates the formation of a ferrite magnetic phase in both synthesis modes. For MA2_RT_750 samples, the formation of ferrite with a high concentration is observed at the stage of sample heating, and with further isothermal synthesis, a slower accumulation of the magnetic phase occurs.

With an increase in the isothermal holding time, *t*, the magnetization of the samples increases and reaches constant values at $t \geq 60$ min. This behavior of ferrite synthesis during electron-beam heating is associated with the acceleration of diffusion processes of interaction between the initial (Fe_2O_3–Li_2CO_3) and transition ($LiFeO_2$) phases due to radiation defect formation and local high-temperature fields created in the samples during their interaction with high-energy electrons.

The results also revealed a higher degree of magnetization of mechanically activated samples (MA1 and MA2 samples in Fig. 5) compared to untreated samples. Moreover, with an increase in the degree of energy intensity of grinding, the content of the magnetic phase in the samples increases. This is due to the fact [31, 32] that mechanical processing leads to a decrease in the particle size and an increase in the contact area between the particles, thereby ensuring a faster ferrite formation.

Thus, thermal analysis showed an accelerated formation of lithium ferrite under conditions of combined use of preliminary mechanical activation of initial reagents mixtures and subsequent heating in a high-energy electron beam at lower temperatures compared to the temperature of traditional synthesis.

4 Conclusion

In this work, the magnetization of lithium ferrite synthesized under conditions of electron-beam and thermal heating of a mechanically activated Fe_2O_3–Li_2CO_3 initial reagents mixture was studied.

It was established that the content of the $Li_{0.5}Fe_{2.5}O_4$ magnetic ferrite phase in the synthesized samples depends on the temperature and time of synthesis, the method of samples heating, as well as on the modes of mechanical grinding. With increasing temperature and synthesis time, the magnetization of the samples increases, which is associated with a large formation of the magnetic phase of lithium ferrite. In this case, the magnetization of samples synthesized in an electron beam is higher than for thermally synthesized samples. This is due to the higher rate of ferrite formation during electron-beam heating.

The degree of powders activity, formed as a result of mechanical grinding of the initial powders in a planetary mill, also affects the rate of lithium ferrite formation in any mode of synthesis. In the samples pre-activated at a high rotation speed of 2220 rpm, intense formation of lithium ferrite occurs at a synthesis temperature of 600 °C, which is significantly lower than the temperature of traditional thermal synthesis, equal to 900 °C.

Acknowledgements This work was supported by the Russian Science Foundation under grant No. 22-19-00183. The authors are grateful to M.V. Korobeynikov (Budker Institute of Nuclear Physics, Novosibirsk) for radiation-thermal experiments using ILU accelerator.

References

1. Abu-Elsaad, N.I., Mazen, S.A., Salem, H.M.: The effect of zinc substitution and heat treatment on microstructural and magnetic properties of Li ferrite nanoparticles. Alloys Compd. **835**, 155227 (2020). https://doi.org/10.1016/j.jallcom.2020.155227
2. Soreto Teixeira, S., Graça, M.P.F., Costa, L.C., Valente, M.A.: Study of the influence of thermal treatment on the magnetic properties of lithium ferrite prepared by wet ball-milling using nitrates as raw material. Mater. Sci. Eng. B **186**, 83–88 (2014). https://doi.org/10.1016/j.mseb.2014.03.008
3. Anupama, A.V., Rathod, V., Jali, V.M., Sahoo, B.: Composition dependent elastic and thermal properties of Li–Zn ferrites. Alloys Compd. **728**, 1091–1100 (2017). https://doi.org/10.1016/j.jallcom.2017.09.099
4. Inoue, A., Kong, F.: Soft magnetic materials, reference module in materials science and materials engineering. Elsevier (2020). https://doi.org/10.1016/B978-0-12-803581-8.11725-4
5. Gao, Y., Wang, Z.: Microwave absorption and electromagnetic interference shielding properties of Li–Zn ferrite-carbon nanotubes composite. J. Magn. Magn. Mater. **528**, 167808 (2021). https://doi.org/10.1016/j.jmmm.2021.167808
6. Sláma, J., Šoka, M., Grusková, A., Dosoudil, R., Jančárik, V., Degmová, J.: Magnetic properties of selected substituted spinel ferrites. J. Magn. Magn. Mater. **326**, 251–256 (2013)
7. Barbara, B.: Louis Néel: His multifaceted seminal work in magnetism. C R Phys. **20**(7–8), 631–649 (2019)
8. Lysenko, E.N., Ghyngazov, S.A., Surzhikov, A.P., Nikolaeva, S.A., Vlasov, V.A.: The influence of ZrO_2 additive on sintering and microstructure of lithium and lithium-titanium-zinc ferrites. Ceram. Int. **45**, 2736–2741 (2019). https://doi.org/10.1016/j.ceramint.2018.09.061
9. Lucas, J.M.F., Prezas, P.R., Soreto Teixeira, S., Ferreira, N.M., Sales, A.J.M., Melo, B.M.G., Graça, M.P.F.: Tuning the magnetic and electric behavior of lithium ferrite using an eco-friendly pectin sol-gel route. J. Sol-Gel Sci. Technol. **98**, 580–592 (2021). https://doi.org/10.1007/s10971-021-05513-1
10. Zahir, R., Chowdhury, F.-U.-Z., Uddin, M.M., Hakim, M.A.: Structural, magnetic and electrical characterization of Cd substituted Mg ferrites synthesized by double sintering technique. J. Magn. Magn. Mater. **410**, 55–62 (2016)
11. Nili-Ahmadababdi, M., Sarraf-Mamoory, R., Yourdkhani, A., Diaconu, A., Rotaru, A.: Magnetic and electrical properties of $Mg_{1-x}Co_xFe_2O_4$ ($x = 0–0.15$) ceramics prepared by the solid-state method. J. Euro. Ceram. Soc. **42**(2), 442–447 (2022)
12. Wang, R., Zhou, T., Zhong, Zh.: Low-temperature processing of LiZn-based ferrite ceramics by co-doping of V_2O_5 and Sb_2O_3: composition, microstructure and magnetic properties. J. Mater. Sci. Technol. **99**, 1–8 (2022)
13. Xu, F., Shi, X., Liao, Y., Li, J., Hu, J.: Investigation of grain growth and magnetic properties of low-sintered LiZnTi ferrite-ceramics. Ceram. Int. **46**, 14669–14673 (2020)
14. Xie, F., Chen, Y., Bai, M., Wang, P.: Co-substituted LiZnTiBi ferrite with equivalent permeability and permittivity for high-frequency miniaturized antenna application. Ceram. Int. **45**, 17915–17919 (2019)
15. Teo, M.L.S., Kong, L.B., Li, Z.W., Lin, G.Q., Gan, Y.B.: Development of magneto-dielectric materials based on Li-ferrite ceramics I. Densification behavior and microstructure development. Alloys Compd **459**, 557–566 (2008)
16. Lysenko, E.N., Nikolaev, E.V., Vlasov, V.A., Surzhikov, A.P.: Microstructure and reactivity of Fe_2O_3–Li_2CO_3–ZnO ferrite system ball-milled in a planetary mill. Thermochim. Acta **664**, 100–107 (2018). https://doi.org/10.1016/j.tca.2018.04.015
17. Berbenni, V., Bruni, G., Milanese, C., Girella, A., Marini, A.: Synthesis and characterization of $LaFeO_3$ powders prepared by a mixed mechanical/thermal processing route. J. Therm. Anal. Calorim. **133**, 413–419 (2018)

18. Lysenko, E.N., Surzhikov, A.P., Vlasov, V.A., Malyshev, A.V., Nikolaev, E.V.: Thermal analysis study of solid-phase synthesis of zinc- and titanium-substituted lithium ferrites from mechanically activated reagents. J. Therm. Anal. Calorim. **122**, 1347–1353 (2015). https://doi.org/10.1007/s10973-015-4849-9
19. Mazen, S.A., Abu-Elsaad, N.I.: Structural, magnetic and electrical properties of the lithium ferrite obtained by ball milling and heat treatment. Appl. Nanosci. **5**, 105–114 (2015)
20. Zhuravlev, V.A., Naiden, E.P., Minin, R.V., Itin, V.I., Suslyaev, V.I., Korovin, E.Yu.: Radiation-thermal synthesis of W-type hexaferrites. IOP Conf. Ser. Mater. Sci. Eng. **81**, 012003 (2015)
21. Kostishyn, V., Isaev, I., Scherbakov, S., Nalogin, A., Belokon, E., Bryazgin, A.: Obtaining anisotropic hexaferrites for the base layers of microstrip SHF devices by the radiation-thermal sintering. Eastern-Euro. J. Enterp. Technol. **5**, 32–39 (2016)
22. Neronov, V.A., Voronin, A.P., Tatarintseva, M.I., Melekhova, T.E., Auslender, V.L.: Sintering under a high-power electron beam. J. Less-Common Metals **117**, 391–394 (1986)
23. Naiden, E.P., Zhuravlev, V.A., Minin, R.V., Suslyaev, V.I., Itin, V.I., Korovin, E.Y.: Structural and magnetic properties of SHS-produced multiphase W-type hexaferrites: influence of radiation-thermal treatment. Int. J. Self Propag. High Temp. Synth. **24**, 148–151 (2015)
24. Malyshev, A.V., Lysenko, E.N., Vlasov, V.A.: Microstructure, electromagnetic and dielectric properties of zinc substituted lithium ferrites prepared by radiation-thermal heating. Ceram. Int. **41**(10), 13671–13675 (2015)
25. Lysenko, E.N., Vlasov, V.A., Surzhikov, A.P.: Investigation of kinetics of lithium ferrite formation under electron beam treatment. Nucl. Instrum. Methods Phys. Res. Sect. B **466**, 31–36 (2020). https://doi.org/10.1016/j.nimb.2020.01.010
26. Berbenni, V., Marini, A., Matteazzi, P., Ricceri, R., Welham, N.J.: Solid-state formation of lithium ferrites from mechanically activated Li_2CO_3–Fe_2O_3 mixtures. J. Eur. Ceram. Soc. **23**, 527–536 (2003). https://doi.org/10.1016/S0955-2219(02)00150-4
27. Bryazgin, A.A.: Electron accelerator for energy up to 5.0 MeV and beam power up to 50 kW with X-ray converter. Probl. Atomic Sci. Technol. **1**, 21–23 (2004)
28. Auslender, V.L., Bryazgin, A.A., Faktorovich, B.L., Gorbunov, V.A., Kokin, E.N., Korobeinikov, M.V., Voronin, L.A.: Accelerators for E-beam and X-ray processing. Radiat. Phys. Chem. **63**, 613–615 (2002)
29. Auslender, V.L.: ILU-type electron accelerator for industrial technologies. Nucl. Instrum. Methods Phys. Res. Sect. B **89**, 46–48 (1994)
30. Lin, D.M., Wang, H.S., Lin, M.L., Lin, M.H., Wu, Y.C.: TG (M) and DTG (M) techniques and some of their applications on material study. J. Therm. Anal. Calorim. **58**, 347–353 (1999)
31. Kavanlooee, M., Hashemi, B., Maleki-Ghaleh, H., Kavanlooee, J.: Effect of annealing on phase evolution, microstructure, and magnetic properties of nanocrystalline ball-milled LiZnTi ferrite. J. Electron. Mater. **41**, 3082–3086 (2012). https://doi.org/10.1007/s11664-012-2235-y
32. Hajalilou, A., Kianvash, A., Lavvafi, H., Shameli, K.: Nanostructured soft magnetic materials synthesized via mechanical alloying: a review. J. Mater. Sci.: Mater. Electron. **29**, 1690–1717 (2018). https://doi.org/10.1007/s10854-017-8082-0

Development of Microprocessor Hardware for Drainage Quality Monitoring by the Level Ground Water

Vladislav Yurchenko, Galina Vavilova, Sergey Bobuyok, Galina Belik, and Pavel Bezkorovainyy

Abstract The analysis of need to control ground water level in urban settings has been carried out. The exterior view of control (piezometric) wells has been represented. The modern methods of ground water level control have been reviewed. The technical requirements for control equipment and level sensors have been developed. The schematic diagrams of the sensor and microprocessor-based control unit have been developed. The main causes of capacitive converters errors are investigated. The main points of the program, field tests methods and the monitoring results at the operating facility are given.

V. Yurchenko (✉)
Department of Information Technology and Security, Abylkas Saginov Karaganda Technical University, 56 Nursultan Nazarbayev Avenue, Karaganda 100027, Republic of Kazakhstan
e-mail: jurchenkovv@mail.ru

G. Vavilova · S. Bobuyok
School of Non-Destructive Testing, National Research Tomsk Polytechnic University, 30 Lenin Avenue, Tomsk 634050, Russia
e-mail: wgw@tpu.ru

S. Bobuyok
e-mail: sab45@tpu.ru

G. Belik
Department of the Technology of Communication systems, Abylkas Saginov Karaganda Technical University, 56 Nursultan Nazarbayev Avenue, Karaganda 100027, Republic of Kazakhstan
e-mail: belicga@mail.ru

P. Bezkorovainyy
Department of Mechanics, Abylkas Saginov Karaganda Technical University, 56 Nursultan Nazarbayev Avenue, Karaganda 100027, Republic of Kazakhstan
e-mail: bpg82_karlion@mail.ru

E. Lysenko et al. (eds.), *Emerging Trends in Materials Research and Manufacturing Processes*, Engineering Materials, https://doi.org/10.1007/978-3-031-38964-1_6

1 Introduction

Settlements with riverbeds within the city suffer from periodic flooding. Drainage works both citywide and individual for separate buildings are used to reduce negative influence. Underground industrial structures for various purposes, deepened premises of residential buildings, underground garages, pedestrian crossings and galleries are also subject to flooding [1]. Flooding leads to the appearance of dampness and fungal infesting of surface and mold, which have a negative impact on buildings and human health.

Other causes of under flooding are precipitation, ground water, surface runoff from surrounding areas, water vapor in ground and rocks. In addition, the impact effect of man-made sources is possible such as leaks from pools, tanks, treatment plants, settling ponds, water pipes, sewerage. These events are exacerbated in restrained urban conditions due to loitering of ground water by underground parts of buildings and structures [2].

Waterproofing systems are traditionally used to protect structures from under flooding [3]. It includes waterproofing membrane, drainage in the form of a sandy prism dump and drainage pipes with filtering mounding of sand and gravel, expanded clay and other materials or wrappers made of artificial fibrous materials. In some cases, a stratum drainage is used represented by layer of crushed stone or gravel with a system of outlet pipe filters [3, 4]. The inspection of the groundwater level depends on the depth of the buildings foundations. Typically this is 0.5–5 m depth from the surface.

To inspection the drainage works carried out by the "KaragandaGIIZ" LLP, the government of Atyrau city (Kazakhstan) set the task of ensuring continuous inspection of the protected buildings in an inexpensive and reliable way for a long period.

2 Methods for Monitoring the Water Level in Wells

After carrying out certain drainage works, it is necessary to inspection groundwater level along the outside perimeter of protected buildings. Inspection is carried out using special wells that are drilled along the building perimeter [5, 6]. Number and distribution of the inspection wells is determined by the building configuration but not less than 4. The well is drilled to a depth of 5 m with a diameter of 100 mm. The well outside appearance is shown in the Fig. 1a.

The water level inspection in wells can be carried out by several methods. The main two are contact and non-contact. Let's consider contact methods. The simplest is the "plopper" method. The plopper is a cavity with an eyelet lowering from a measuring cord [7]. When it touches the water surface, it makes a distinctive plopping sound.

Electrocontact level gauges [8, 9] are used only while using steel casing pipes. Most of old-design level gauges (which are still used today) have the following

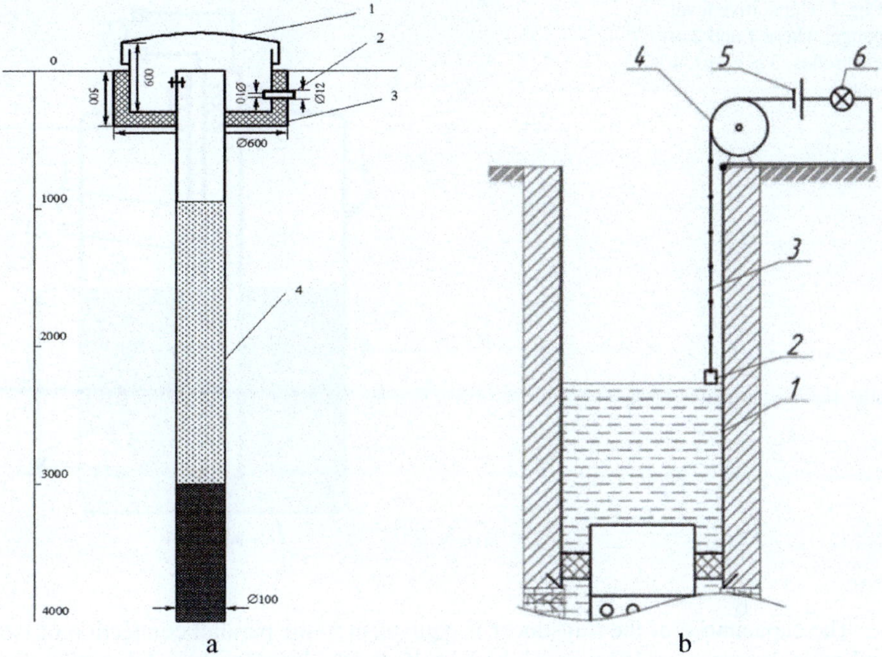

Fig. 1 Inspection well (**a**), where 1 is cover, 2 is tube for cable entry, 3 is body, 4 is well casing and Electrocontact level gauges (**b**), where 1 is casing string; 2 is contact bar; 3 is single-core cable with a measuring marks; 4 is tape ruler; 5 is battery; 6 is bulb

principle of operation: one wire is lowered into the well, the second is connected to a metal casing pipe, a light bulb or a pointer indicator serves as an indicator of contact with water (electrical circuit is closed) [10]. Therefore in a rainy weather it is impossible to provide measurements. It is also worth noting that the casing pipes should be metal only. The measuring scheme shown in Fig. 1b [11].

The MPM416WK submersible hydrostatic level sensor has a protected steel pipe structure and is flange-mounted. Purpose—continuous measurements in containers up to 4 m deep. Operating principle is based on measuring change of hydrostatic pressure on the inner diaphragm. Signal releases from strain gauge transformer and fed to the microprocessor analog input. The device allows monitoring the water level and writing data directly to the "paper" logbook. The presence of analog output signal allows integrating level gauge into a wireless well monitoring and controlling system [12].

The cost of such devices is quite high, therefore, a capacitive water level sensor is offered, which is not expensive and does not require additional maintenance during long-term operation [13].

The transducer in these level gauges is two parallel electrodes immersed in a tank [14, 15]. Figure 2 show a diagram of a level gauge with cylindrical electrodes.

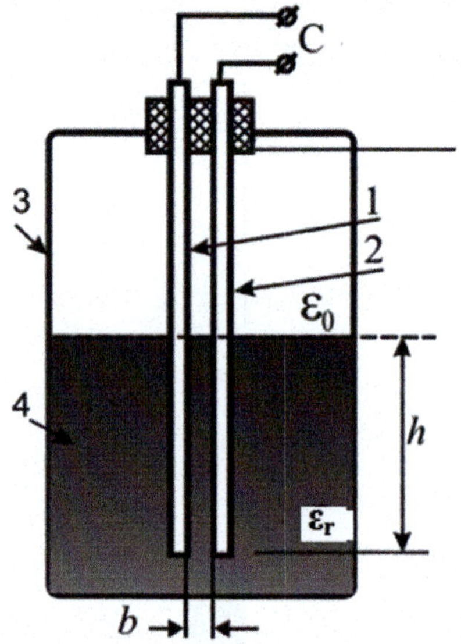

Fig. 2 Capacitive level gauge, where 1 and 2 are electrodes, 3 is corps, 4 is water

The capacitance of the transducer is equivalent to the parallel connection of two cylindrical capacitors [16], one of which is filled with a liquid with a relative dielectric constant ε_r and has a height *h*, the other has a height *H–h* and is free of liquid [17]. The transducer capacitance will be equal

$$C = C_0(\varepsilon_r h + H - h) = C_0[H + h(\varepsilon_r - 1)], \tag{1}$$

where C_0 is capacitance of the transducer length without liquid, *H* is electrodes height, *h* is liquid level height. Input quantity of the transducer is $h(\varepsilon_r - 1)$

As an example, the DLM-35 capacitive level sensors are presented [18] manufactured in the Russian Federation. DLM-35 is designed for constant measurement of level height by liquid and bulk substances in containers. A feature of the brand is the possibility of using DLM-35 capacitive level sensors in contaminated and clean liquids, at high temperatures and in explosive environments.

Technical specification contents next characteristics: power voltage is 9–34 V (DC); current output is 4–20 μA (2 wires); voltage output is 0–10 V (3 wires); Nonlinearityis max 1%; temperature error is <0.05% K; voltage error (for voltage and current output) has maximum of 0.3 μA/V and 0.1 mV/V; ambient temperature is 40 °C ··· + 85 °C; process temperature is up to 200 °C [12, 19].

Thus, a simple and reliable method is the capacitive method for measuring the ground level in special wells. It is necessary to use a microprocessor controller to collect and store information. Here are the basic requirements for the hardware.

3 The Prototype Manufacturing and Field Tests

The microprocessor-based ground water level inspection equipment should consist of level sensors (LS) and a control unit (CU).

Operating conditions should comply with the following parameters:

- An ambient temperature from −40 to +40 °C;
- A relative humidity up to 98% at temperature of +35 °C [20].

Main functions performed:

- The water level inspection in four wells located along the perimeter of building;
- The transmission of level data upon request from CU;
- The interrogation of the well level sensor after 24 h;
- The storage of information for at least 30 days;
- The data transmission to the operator using manual retrieval [21].

The control unit should have a power supply unit and a microprocessor unit with control buttons and an indication. The Microprocessor unit should be a printed circuit board with elements of an electrical circuit diagram and control buttons located on it: “Date No.”, “Measurement No”, “Data reset” and “Data view”.

There should be indicators on the board:

- “Date No”. Must be 1 digit;
- “Measurement No”. Must be 2 digits;
- “Indications, m”. Must be 3 digits.

The power supply should be a board with circuit elements located on it which must ensure the conversion of 220 V AC voltages to DC +5 V and ±12 V DC voltages to a power the microprocessor unit [19, 22].

Level sensors should be a capacitive sensor consisting of an electrode and an electronic unit. Sensor electrodes should be made of anti-corrosion material for a ground water, up to 5 m long.

The electronic unit of the level sensor should be powered from the control unit CU with a DC voltage. The sensor configuration should be made in accordance with well dimensions shown in Fig. 1a [23].

The prototypes were developed and manufactured in the conditions of scientific and production “KAYUR” LLP, Karaganda and tested in the city of Atyrau (Kazakhstan) to control level of a drainage water near building of the Culture Palace.

The block diagram of the microprocessor control unit CU is shown in Fig. 3. The main element is a microprocessor controller based on PIC16F876 from Microchip, as well as a circuit breaker to protect the circuit from overload and overvoltage and a power supply [24].

The control unit is a microprocessor device that collects information from level sensors, number of which should be no more than four [25].

Analog signals from the level sensors are fed first to the input of the current signal converter into a voltage signal with a value, and then to the input of the ADC

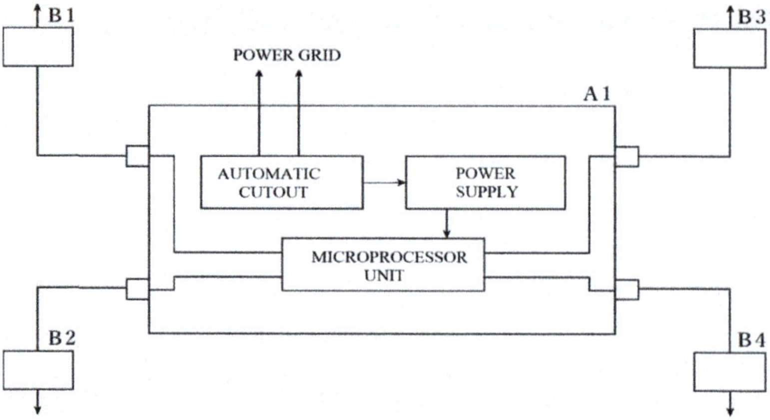

Fig. 3 Equipment block diagram, where A1 is microprocessor control unit, B1 ÷ B4 are capacitive level sensors

controller. An indication unit is used to display information about the well number and its level. There are control buttons "Date No", "Measurement No", "Data reset" and "Data view" on the board.

Consider the errors of capacitive converters. The main one is the temperature error. Dissimilar sensor materials have different linear expansion coefficients. Temperature *t* change can lead to change in the distance between the electrodes [26, 27]. This error can be significant with large electrode sizes. It decreases when the transducers are in differential connection [28].

The test methodology main points are given below [14, 26].

1. Testing for environmental influences is carried out during trial operation by periodic observations of the nodes state located inside shells [29].
2. Checking distances of LS from CU is carried out in a laboratory conditions. It is necessary to connect the LS to the CU through resistances, values of which are determined by the used cable parameters. The core resistance of telephone mine cable is 37.7 Ω/km. Thus, the values of additional resistances will be 19 Ω. A functional check is carried out according to Sect. 3 after connecting resistances [30].
3. Checking the controlled depth are carried out on a special stand or during field tests. In the absence of water (the water level is below the electrode), a 5 mA signal should be output from the level sensor. As the water level rises, up to the wellhead, the output signal should decrease proportionally down to 0 mA [31].
4. The absolute error is checked by any device or method that has an appropriate accuracy class and does not affect the readings of the remote control sensor during the tests according to clause 3 [32].
5. Checking the temperature error is carried out during the equipment experimental operation by measuring the water level in the well and comparing the data obtained with the equipment indications [33, 34].

6. Checking the absolute error when changing the supply voltage are as follows. Apply 242 V supply voltage and carry out measurements according. Apply voltage 187 V and re-measure according [35–37].
7. Checking the readout and reset of information are carried out by the control unit with connected level sensors (field tests) or simulators during laboratory tests. To do this, the "Reset" button should be pressed and released. The sensor readings should be recorded. Press the buttons "Sensor No." and "Information output", after which the corresponding information is displayed on the display unit [38, 39].
8. Checking the information storage for 30 days is carried. The verification is carried out during full-scale tests and trial operation.
9. Testing the reliability of the equipment are carried out during trial operation.

4 Test Results

Tests of the equipment prototype were carried out in accordance with program and methodology of predelivery test in the conditions of the Atyrau city to inspection the ground water level after drainage works to protect of the Culture Palace.

The control unit was installed in the premises of the Culture Palace. Level sensors were installed in the wells and connected to the control unit with a 4-core cable of TSMSC (telephone subscriber mine with steel core cable) 1 × 4 type.

The ambient temperature during the tests met the requirements and was 32 °C. The output current from the level sensors was measured, value of which is directly proportional to the value of ground water depth in meters. The test results are shown in Table 1.

The current level corresponded to the values of display unit. The error did not exceed 10 cm. The duration of the tests is 5 days.

Table 1 Test results

	Well No. 1			Well No. 2			Well No. 3			Well No. 4		
	I_{ls} (mA)	L [m]	Δ	I_{ls} (mA)	L (m)	Δ	I_{ls} (mA)	L (m)	Δ	I_{ls} (mA)	L (m)	Δ
1	1.35	1.35	0	1.10	1.10	0	1.00	1.00	0	0.93	0.93	0
2	1.34	1.34	0	1.08	1.08	0	0.99	0.98	0	0.87	0.87	0
3	1.34	1.33	0	1.08	1.08	0	0.99	0.98	0	0.85	0.85	0
4	1.32	1.31	+0.1	1.07	1.06	+0.1	0.98	0.98	+0.1	0.80	0.80	0
5	1.31	1.31	0	1.05	1.04	+0.1	0.96	0.97	+0.1	0.77	0.78	−0.1

Notes
I_{ls} is current at the level sensor output of AWCL equipment
L is water level in the well at the time of measurement
Δ is error of correspondence between the depth to the water surface in the well and the output current

Measurements were taken daily every 24 h. Control measurements of the ground water level in the wells were carried out by the "plopper" method, which consists in lowering device into the wells and after a sound signal it raised to the top and the distance from the surface to the water was read. These values were compared with the current value and indications on the CU. Discrepancies are within the specified error.

The prototype has been tested and recommended for the manufacture of a pilot batch. The prototype was left for trial operation for one year. During the tests, climatic temperatures varied from minus 23 °C in winter to +35 °C in summer. Deviations and refusals were not observed.

5 Conclusion

The paper presents the equipment used to monitor groundwater level in urban settings. This system allows continuous control during extended period. Circuit engineering of the equipment is based on modern methods of groundwater level monitoring.

The paper presents requirements for this equipment and shows its main technical characteristics in detail. As a result, prototype equipment was made. The article shows its structure and circuit diagram and the principle of its operation is described in detail.

The prototype equipment passes field tests. As a result of which equipment working capability was checked, metrological characteristics were evaluated. The article presents some results obtained in the approbation process (Table 1).

As a result of tests equipment proves to be inexpensive and reliable, allowing work in urban settings during a long time in various temperature and climatic conditions.

This equipment is recommended for producing pilot batch of 10 sets for installation in Atyrau and Nur-Sultan for monitoring the groundwater level near buildings after drainage work.

References

1. Akadiri, P.O., Chinyio, E.A., Olomolaiye, P.O.: Design of a sustainable building: a conceptual framework for implementing sustainability in the building sector. Buildings **2**(2), 126–152 (2012). https://doi.org/10.3390/buildings2020126
2. Alsalme, A., Al-Zaqri, N., Ullah, R., Yaqub, S.: Approximation of ground water quality for microbial and chemical contamination. Saudi J. Biol. Sci. **28**(3), 1757–1762 (2021). https://doi.org/10.1016/j.sjbs.2020.12.017
3. Fathollahi, A., Coupe, S.J.: Life cycle assessment (LCA) and life cycle costing (LCC) of road drainage systems for sustainability evaluation: quantifying the contribution of different life cycle phases. Sci. Total Environ. **776**, 145937 (2021). https://doi.org/10.1016/j.scitotenv.2021.145937
4. Nizhegorodov, A.I., Gavrilin, A.N., Moyzes, B.B., Cherkasov, A.I., Zharkevich, O.M., Zhetessova, G.S., Savelyeva, N.A.: Radial-piston pump for drive of test machines. IOP Conf.

Ser.: Mater. Sci. Eng. **289**(112), 0120146 (2018). https://doi.org/10.1088/1757-899X/289/1/012014
5. Gokdemir, C., Rubin, Yo., Xiaojun, L., Yandong, L., Hao, X.: Vulnerability analysis method of vegetation due to groundwater table drawdown induced by tunnel drainage. Adv. Water Resour. **133**, 103406 (2019). https://doi.org/10.1016/j.advwatres.2019.103406
6. Salo, H., Mellin, I., Sikkilä, M., Nurminen, J., Äijö, H., Paasonen-Kivekäs, M., Virtanen, S., Koivusalo, H.: Performance of subsurface drainage implemented with trencher and trenchless machineries, Agricul. Water Manag. **213,** 957–967 (2019). https://doi.org/10.1016/j.agwat.2018.12.010
7. Gavrilin, A.N., Chuprin, E.A., Moyzes, B.B., Halabuzar, E.A.: Land-based sources of seismic signals. In: Proceedings of 2014 International Conference on Mechanical Engineering, Automation and Control Systems, MEACS 2014 6986947 (2014). https://doi.org/10.1109/MEACS.2014.6986947
8. Czaja, Z.: A measurement method for lossy capacitive relative humidity sensors based on a direct sensor-to-microcontroller interface circuit. Measurement **170**, 108702 (2021). https://doi.org/10.1016/j.measurement.2020.108702
9. Paczesny, D., Tarapata, G., Michał, M., Jachowicz, R.: The capacitive sensor for liquid level measurement made with ink-jet printing technology. Procedia Eng. **120**, 731–735 (2015). https://doi.org/10.1016/j.proeng.2015.08.776
10. Sergeyev, V., Yurchenko, V., Ayzhambayeva, S., Vavilova, G., Belik, M., Serebryakov, S.: Researches of air and fuel rate influence on oxygen level in emissions of new type medium power coal boiler. IOP Conf. Ser.: Mater. Sci. Eng. **457**(1), 012023 (2018). https://doi.org/10.1088/1757-899X/457/1/012023
11. Raneev, G.G.: Methods and means of measurement, Moskow (2004)
12. 5 benefits of conductive over ultrasonic sensors for water utilities level measurement. Information. https://www.gillsc.com/applications/sludge-level-measurement
13. David, A.: Bell Electronic Instrumentation and Measurements, Ontario (2007)
14. Yurchenko, V.V., Belik, M.N.: Analogue measuring devices. Karaganda (2016)
15. Dinh, B.H., Go, G.H., Kim, Y.S.: Performance of a horizontal heat exchanger for ground heat pump system: effects of groundwater level drop with soil–water thermal characteristics. Appl. Therm. Eng. **195**, 117203 (2021). https://doi.org/10.1016/j.applthermaleng.2021.117203
16. Goldshtein, A., Vavilova, G., Rumkin, A., Starý, O.: Detection of insulation defects in automated in-process testing of electric wire during its extrusion. Mater. Sci. Forum **942**, 141–150 (2019). https://doi.org/10.4028/www.scientific.net/MSF.942.141
17. Shahid, M., Laxmeesha, S., Meraj, A., Tarikul, I., Maryam, S.: An accurate digital converter for lossy capacitive sensors. Sens. Actuators A **331**, 112958 (2021). https://doi.org/10.1016/j.sna.2021.112958
18. Hongliang, Q., Pega, H.: Design and calibration of capacitive sensors for measuring void fraction in vertical headers of microchannel heat exchangers. Int. J. Refrig **129**, 224–236 (2021). https://doi.org/10.1016/j.ijrefrig.2021.05.012
19. Sanchez, J., Canton, M.P.: Microcontroller Programming. The Microchip PIC Taylor & Francis Group (2006)
20. Lysenko, E., Nikolaev, E., Vlasov, V., Surzhikov, A.: Microstructure and reactivity of Fe_2O_3–Li_2CO_3–ZnO ferrite system ball-milled in a planetary mill. Thermochim. Acta **664**, 100–107 (2018). https://doi.org/10.1016/j.tca.2018.04.015
21. Zia Ur Rahman, M., Aldossary, M., Tarikul, I.A.: Constant phase impedance sensor for measuring conducting liquid level. ISA Trans. **115**, 250–258 (2021). https://doi.org/10.1016/j.isatra.2021.01.024
22. Lysenko, E.N., Malyshev, A.V., Vlasov, V.A., Nikolaev, E.V., Surzhikov, A.P.: Microstructure and thermal analysis of lithium ferrite pre-milled in a high-energy ball mill. J. Therm. Anal. Calorim. **134**(1), 127–133 (2018). https://doi.org/10.1007/s10973-018-7549-4
23. Yang, X., Wang, Y., Qing, X.: A flexible capacitive sensor based on the electrospun PVDF nanofiber membrane with carbon nanotubes. Sens. Actuators A **299**, 111579 (2019). https://doi.org/10.1016/j.sna.2019.111579

24. Fu, Y., Li, W., Chen, S., Wang, Z., Liu, Y.: Experimental and simulation study on the capacitive liquid level gauge of LNG vehicular cylinder. Measurement **164**, 107930 (2020). https://doi.org/10.1016/j.measurement.2020.107930
25. Hanni, J.R., Venkata, S.K.: Does the existing liquid level measurement system cater the requirement of future generation? Measurement **156**, 107594 (2020). https://doi.org/10.1016/j.measurement.2020.107594
26. Sedha, R.S.: Electronic measurement and instrumentation. S Chand & Co Ltd, (1996)
27. Wang, J., Mao, J., Han, X., Li, Y.: Study on analytical solution model of heat transfer of ground heat exchanger in the protection engineering structure. Renew. Energy **179**, 998–1008 (2021). https://doi.org/10.1016/j.renene.2021.07.081
28. Moropoulou, A., Labropoulos, K.C., Delegou, E.T., Karoglou, M., Bakolas, A.: Non-destructive techniques as a tool for the protection of built cultural heritage. Constr. Build. Mater. **48**, 1222–1239 (2013). https://doi.org/10.1016/j.conbuildmat.2013.03.044
29. Mataei, B., Nejad, F.M., Zahedi, M., Zakeri, H.: Evaluation of pavement surface drainage using an automated image acquisition and processing system. Autom. Constr. **86**, 240–255 (2018). https://doi.org/10.1016/j.autcon.2017.11.010
30. Shen, D., Chen, K.P.: Symmetric drainage flow of a compressible fluid from a fracture: analytical solution and slip-like flow rate. Mech. Res. Commun. **99**, 15–21 (2019). https://doi.org/10.1016/j.mechrescom.2019.06.002
31. Horstmann, S., Henderson, C.J., Hall, E.A., Daly, R.: Capacitive touchscreen sensing—A measure of electrolyte conductivity. Sens. Actuators B Chem. **345**, 130318 (2021). https://doi.org/10.1016/j.snb.2021.130318
32. Mondal, P.P., Mahapatra, P.L., Das, S., Saha, D.: Study on the novel capacitive moisture sensing behaviour of nickel chromite nanoparticle based thick film. Measurement **163**, 107992 (2020). https://doi.org/10.1016/j.measurement.2020.107992
33. Surzhikov, A.P., Frangulyan, T.S., Ghyngazov, S.A., Lysenko, E.N.: Investigation of oxidation processes in non-stoichiometric lithium-titanium ferrites using TG analysis. J. Therm. Anal. Calorim. **102**(3), 883–887 (2010). https://doi.org/10.1007/s10973-010-0912-8
34. Czaja, Z.: A measurement method for capacitive sensors based on a versatile direct sensor-to-microcontroller interface circuit. Measurement **155**, 107547 (2020). https://doi.org/10.1016/j.measurement.2020.107547
35. Chung, C.K., Ku, C.A., Wu, Z.E.: A high-and-rapid-response capacitive humidity sensor of nanoporous anodic alumina by one-step anodizing commercial 1050 aluminum alloy and its enhancement mechanism. Sens. Actuators B Chem. **343**, 130156 (2021). https://doi.org/10.1016/j.snb.2021.130156
36. Surya, S.G., Yuvaraja, S., Varrla, E., Baghini, M.S., Palaparthy, V.S., Salama, K.N.: An in-field integrated capacitive sensor for rapid detection and quantification of soil moisture. Sens. Actuators B Chem. **321**, 128542 (2020). https://doi.org/10.1016/j.snb.2020.128542
37. Rahim, I., Shah, M., Khan, A., Luo, J., Zhong, A., Li, M., Ahmed, R., Li, H., Wei, Q., Fu, Y.: Capacitive and resistive response of humidity sensors based on graphene decorated by PMMA and silver nanoparticles. Sens. Actuators B Chem. **267**, 42–50 (2018). https://doi.org/10.1016/j.snb.2018.03.069
38. Surzhikov, A.P., Lysenko, E.N., Vasendina, E.A., Sokolovskii, A.N., Vlasov, V.A., Pritulov, A.M.: Thermogravimetric investigation of the effect of annealing conditions on the soft ferrite phase homogeneity. J. Therm. Anal. Calorim. **104**(2), 613–617 (2011). https://doi.org/10.1007/s10973-010-1074-4
39. Surzhikov, A.P., Frangulyan, T.S., Ghyngazov, S.A., Lisenko, E.N., Galtseva, O.V.: Physics of magnetic phenomena: investigation of electroconductivity of lithium pentaferrite. Russ. Phys. J. **49**(5), 506–510 (2006)

The Study of the Parameters of Amplitude-Modulated Sweep Signal of the Shock Vibration Source of Seismic Signals

Boris Moyzes and Anatolij Nizhegorodov

Abstract The purpose of the research is to document the development process of a low-weight seismic shock vibration sensor with an amplitude-modulated sweep signal, and which possess good signal-to-noise ratio characteristics. The development of such a source of seismic signals will make it possible to study the interior of the Earth in difficult terrain geographic regions. To achieve this goal, the designs of seismic sources were analyzed and their "positive" design features were defined; the effect of different law of amplitude-modulated sweep signal on the degree of compactness of the autocorrelation function was studied; theoretical studies were carried out on the oscillatory system seismic source behavior with the simultaneous impact of a force pulse from a falling load, which forms the pressing force of the base plate and a sweep signal from the oscillation generator; the experimental studies were carried out on a vibration bench which is a physical model of the source; the possibility of excitation of a sweep signal with amplitude modulation was studied; the parameters of the excitation process and the operation modes of the vibration mechanism for the formation of a useful vibration load are recommended.

1 Introduction

The great attention is paid to the issues of vibration studying in various fields of activity [1–10]. There is a vibration in almost any technological system. The vibration makes worse the quality of a technological system work [11–23].

B. Moyzes (✉)
School of Non-Destructive Testing, National Research Tomsk Polytechnic University, 30 Lenin Avenue, Tomsk 634050, Russia
e-mail: mbb@tpu.ru

A. Nizhegorodov
Department of Construction and Road-Making Machines and Hydraulic Systems, Irkutsk National Research Technical University, 83 Lermontov Street, Irkutsk 664074, Russia
e-mail: nastromo_irkutsk@mail.ru

E. Lysenko et al. (eds.), *Emerging Trends in Materials Research and Manufacturing Processes*, Engineering Materials, https://doi.org/10.1007/978-3-031-38964-1_7

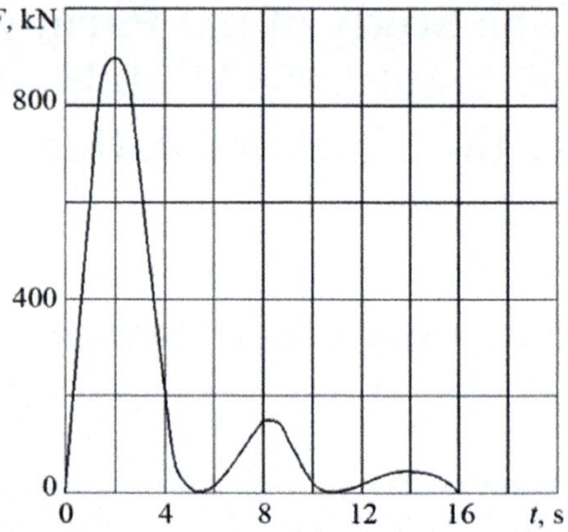

Fig. 1 An impulse action example

A wide range of vibration and impact machinery is used for various purposes in modern technology, including for geophysical seismic exploration of the Earth's geological environments.

Due to variety of tasks facing seismic exploration a whole range of seismic sources of various actions (vibration, code-pulse, impulse are widely used). The use of a certain type of source is determined by its advantages and disadvantages.

First type sources generate and transmit a probing signal (load) to the ground in the form of a sweep signal. Sweep signal is a signal with frequency modulation (Fig. 1), which possess good signal-to-noise ratio characteristics and high stability of its excitation process.

At the same time, a large clamping mass is required, to form and transfer to the ground a large amplitude load, which significantly increases the weight of the entire machine and limits the use of vibration seismic sources in difficult terrain.

Therefore, the blasting works or works with impulse mechanisms are preferred in these regions. Poor stability of the probing signal excitation process is a part of specific performance for these works, due to the influence of an inhomogeneous physical environment near the charge. The blasting is harmful to the environment as well.

The designers of new technology are constantly looking for the ways to create a source of seismic signals of low weight, which can be used in as many geographical areas as possible [24–26].

The requirements for the excited load and the weight of the entire machine are compulsory and stipulated by.

Seismic signal sources are being developed at the National Research Tomsk Polytechnic University based on:

- hydrostatic vibration mechanisms with periodically falling load on the actuator;
- actuators, consisting of a shock absorber on an intermediate mass;
- intermediate mass on preloaded elastic non-linear shells (high-pressure hoses).

A sweep signal is excited by a high-frequency generator in the high-pressure hoses (HPH) when the falling weight interacts with the shock absorber.

The non-linear stiffness characteristic HPH, through which the generated load is transferred to the ground, allows you to create large useful dynamic forces.

The author has resolved a number of problems which will make it possible to create a lighter mechanism for seismic exploration with an existing once, and a signal which possess a good signal-to-noise ratio properties, namely:

- to determine the "positive" design features of pulsed, code-pulsed and vibrating seismic sources their designs were analyzed so that the "positive" design features must be taken into account when designing a new device;
- the influence of various forms of the amplitude envelope of the sweep signal oscillations (the law of variation of the oscillation amplitude) on the results of correlation analysis (the degree of compactness of the autocorrelation function) was studied;
- theoretical studies of the oscillatory system behavior of the seismic source were carried out under the simultaneous impact on the oscillatory system of a force impulse from a falling load and a sweep signal, which results in the formation of a load (sweep signal with amplitude modulation) which is transmitted to the ground;
- the experimental studies were carried out by means of the vibrational mechanism physical simulator, where the load falling process is simulated by an additional generator of low-frequency oscillations operating at the frequency of the main resonance;
- it was studied the possibility of excitation of a sweep signal with the amplitude modulation formed by the carrier frequency of the force pulse and the frequency of the sweep signal. To determine the initial sweep frequency of the sweep signal, at which this influence is the least, the influence of the frequencies of the sweep signal generator on the frequency of the force pulse was estimated;
- the parameters of the excitation process and the modes of the vibration mechanism operation to form of a useful vibration load were recommended.

2 Analysis of Seismic Signal Sources

In seismic exploration, the effective solution of problems depends on the technical characteristics of the sources used and on the parameters of the probing signal excited by them. A number of requirements are imposed on sources and the signal emitted by them, the main of which are:

- the smallest weight of a device that emits a signal with greater power and information content, which determines the degree of better separation of the recorded signal against the background of received interference;

- the permanence of the excited signal properties at different points during repeated emissions, which makes it possible to accumulate effects to reduce the level of interference.

Nowadays there are many seismic ground sources that have a number of both advantages over each other and disadvantages depending on the energy converted in the emitters, the nature of the excited load, the actuator, the type of drive.

To eliminate existing shortcomings by making changes to existing designs or by creating a new design of the device the work on the designing of new equipment is carried out. For this purpose, the applied seismic ground sources are analyzed and described below. According to the nature of the excited load, the sources can be divided into three classes: pulsed (the force is formed in the form of individual pulses), vibrational (in the form of a certain periodic function) and code-pulse (individual pulses following one after another at certain intervals).

2.1 Impulse and Impulse-Code Sources of Seismic Vibrations

It is possible to form probing signals in the form of pulse due to ability to dose the excitation energy and no need to correct the energy and phase of the emitted signals.

There are many methods of pulses excitation. The methods which have received the greatest practical application is discussed below.

Gas-dynamic sources of seismic signals use the detonation energy of the gas mixture in the chamber. The magnitude of the impulse transmitted to the ground depends on the volume of the gas chamber and the initial pressure of the operating mixture. To increase the exploration depth, an increase in the volume of the chamber is required. This leads to a significant increase in the weight parameter of the machine. When using these sources to obtain the required signal-to-noise ratios, it is provided to use about a hundred sources. To accumulate effects several sources operate synchronously.

Seismic impact sources use the energy of a falling load. The value of the energy depends on the weight of the load and the height of its rise. The process of forming the magnitude of the load on the ground is characterized by low stability, i.e. the repeatability of the pulse parameters.

Electromagnetic sources of seismic signals have an electromagnetic hammer in design. It consists of concentric windings of forward and reverse motion. The armature inside the cavity of the windings, when switched on alternately, performs a reciprocating movement and forms an impulse load on the hammer blade, which is transferred to the ground. These devices are simple and reliable, provide a stable forming of significant force effects with the ability to control them in frequency and amplitude (Fig. 1).

Thus, pulse mechanisms are characterized by common advantages such as a simple design and significant dynamic forces, a common disadvantage such as a poor signal-to-noise ratio characteristics. The disadvantage causes a weak ability to extract useful information against the background of interference from other operating machines, in connection with construction, blasting, etc.

The particular interest for consideration is pile driving equipment. It can be created the powerful sources of seismic vibrations on it base.

According to the nature of the excited load, the piling machines are divided into two main classes:

- diesel-hammers are the piling machines when the piles are immersed due to the shock force impulse;
- vibratory hammers are the piling machines when piles are driven by impact force impulse and additional oscillatory load at the moment of impulse action (Fig. 2a). At certain frequencies of the vibrator 2, there is a complete failure of the pile from the ground. After a breakdown, the resistance of the ground to translator movement decreases.

When replacing the pile head 4 and the pile 5 in these mechanisms by the base plate 10, which is installed on the ground at the place where the signal is transmitted to the ground, we will get a seismic source that emits an impulse load.

The disadvantage of these sources will be:

- the changed conditions of signal transmission to the ground due to the gradual compaction of the ground;

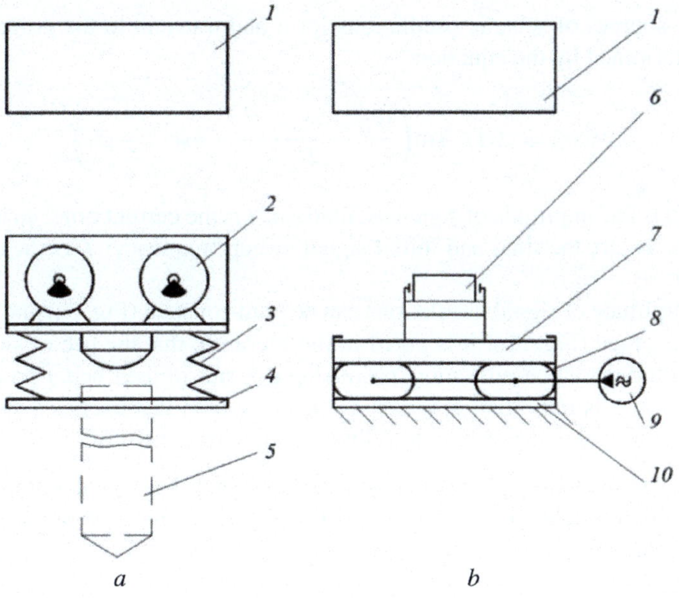

Fig. 2 The formation scheme of vibro-impact load, where **a** is the vibrating hammer, **b** is the proposed scheme of vibratory mechanism, where 1 is the load; 2 is the vibrator; 3 is the springs; 4 is the pile head; 5 is the pile; 6 is the shock absorber; 7 is the intermediate plate; 8 is the elastic shells (HPH); 9 is the oscillation generator; 10 is the base plate

- poor visibility of individual pulses near construction, road works, etc.

In this case, in the second device, each individual pulse will include additional oscillations. It is necessary to form a certain law of fluctuation change to isolate each pulse during the processing of the registered information.

Replacing the mechanical springs 3 by elastic shells 8, previously installed tne shock absorber 6 on the intermediate mass 7 to control the shape of the oscillatory load envelope, well get the scheme being developed in the work.

Thus, if, at the moment of impact, generator 9 excites an oscillatory load with, for example, a linearly increasing frequency in HPH 8, having previously installed shock absorber 6 on intermediate mass 7 to control the shape of the oscillatory load envelope, we'll get the developed scheme (Fig. 2b). This scheme includes:

- the clamping system based on a falling load;
- vibratory mechanism with elastic shells (HPH).

Various sequences of impulses, i.e. impulses at regular intervals are used for Earth sounding in impulse-coded seismic. The pulse signal coding makes it possible to increase the signal-to-noise ratio characteristics.

2.2 *Vibration Sources of Seismic Oscillations*

Vibratory sources of seismic oscillations form and transmit to the ground a sweep signal determined by the equation

$$F(t) = A(t) \cdot \sin\left(\frac{(\omega_f - \omega_0) \cdot t^2}{2T} + \omega_0 \cdot t + \varphi_0\right), \tag{1}$$

where $A(t)$ is the amplitude of signal oscillations; t is the current time; φ_0 is the initial phase; ω_f, ω_0 are the final and initial signal sweep frequency; T is the duration of the signal.

The amplitude of signal oscillations can be both constant (Fig. 3a) and have some smooth envelope (Fig. 3b). It is noted in the literature that the use of sweep signal with amplitude modulation is more promising in terms of their better processability.

Therefore, it is important to create a seismic source that excites a sweep signal with a smoothly varying amplitude. Vibration sources consist of:

- a control system and a system that generates a signal. The signal is amplified on an electro-hydraulic converter and transmitted by an actuator vibration mechanism to the ground;
- an energy source;
- an actuator that converts the incoming energy into the transferring of a moving element;

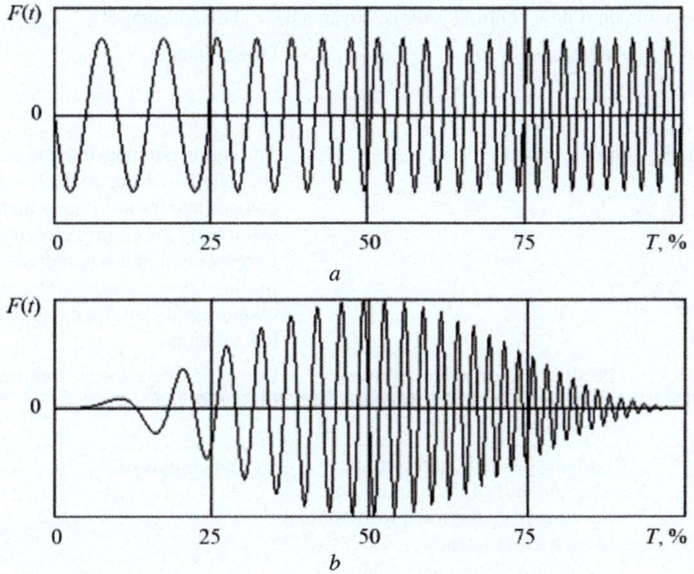

Fig. 3 Sweep signal, where **a** is sweep signal with constant amplitude, **b** is sweep signal with variable amplitude

- a system that provides that the base plate of the source is pressed against the ground at the signal emission moment.

At the same time, a variety of actuators and control methods makes it possible to create mechanical and hydraulic seismic vibration sources on their basis.

The first emitters are created on the basis of rotary electromechanical machines. The vibration load is formed due to rotating unbalances in these machines. Adjustment to a given oscillation frequency is performed by an electric motor. Vibration emitters of this class are characterized by relatively low energy consumption, separate control of the frequency and amplitude of the vibration load (probing signal). The disadvantage is the excitation of a signal with a rectangular envelope.

Hydraulic radiators work on the basis of a variable fluid flow created by throttle and voluminous regulation.

Open-loop and servo (tracking) throttle control systems are used in seismic sources (Table 1). An analysis of the designs of technical systems with throttle control revealed their common drawback (low efficiency at high energy costs).

In hydraulic vibration mechanisms with volume control, a variable load law is formed by an oscillation generator without fluid throttling, i.e. without additional energy loss [27–29]. To convert a variable fluid flow into the transmition of a movable body that forms a load on the ground, the following are used:

- hydraulic cylinders. Their use is limited by the service life due to wear of the seals and possible leakage of the working fluid between the cavities;

Table 1 Variable fluid flow control systems: advantages and disadvantages

Regulation type	Advantages	Disadvantages
Throttle		
• Open-loop	Simple designe	The significant non-linearity of the speed characteristic leads to the need to compensate for non-linear distortions using complex control circuits Large weight and size indicators, machine power and energy costs to obtain significant forces on the ground, low efficiency
• Servo (tracking)	The ability to generate arbitrary waveforms with little distortion and compactness	Low efficiency, design complexity, limited frequency range
Voluminous	Rigid mechanical characteristic (independence of the oscillation amplitude from the force load on the piston) high efficiency Low metal consumption when creating high working pressures (significant loads on the ground)	Design complexity

- HPH (elastic shells) adapted to work at high pressures and at high frequencies (Fig. 4).

Thus, the prospects for application of seismic sources are obvious:

- vibrational sources are more informative;

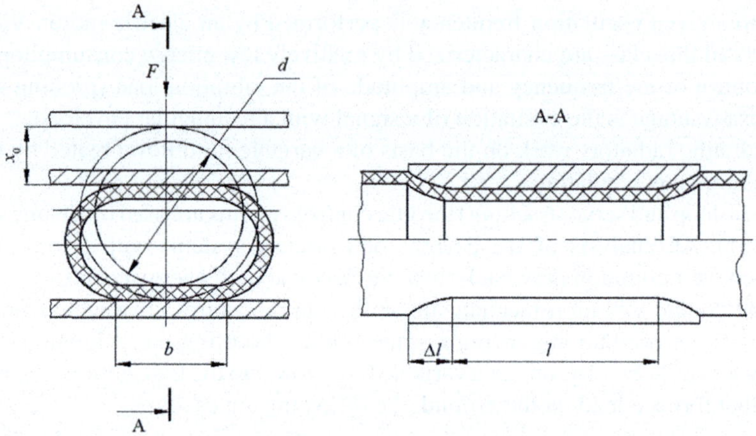

Fig. 4 HPH, where x_0 is the value of the preliminary radial deformation by the force F from the inertial mass; l, b are the length and width of the preloaded section by the inertial mass

- hydraulic sources make it is possible to excite amplitude-modulated sweep signals;
- the less energy consumption is due to a volumetric method of regulation;
- on HPH (the formation of greater ground loads, i.e. greater exploration depth, and a larger frequency range).

The concept of an actuator of a seismic source designing was formed. It consists of:

- an oscillation generator with four working cavities (Fig. 5) connected either in pairs with the upper and lower HPH (Fig. 6) or crosswise (Fig. 7). In this case, the amount of vibration on the intermediate plate x_2 and on the base x_3 will depend on the amount of fluid volume dW supplied to the HPH, which depends on the angle of inclination of the disk γ_d and the maximum stroke of the pistons x_1;
- preliminary preloaded by inertial mass HPH to form the working contact area

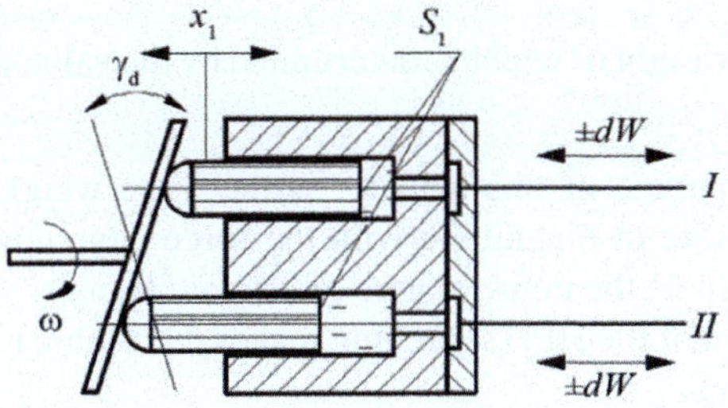

Fig. 5 The design scheme of the oscillation generator, where x_1 is the current coordinate of the piston displacement; S_1 is the piston area; γ_d is the disk inclination angle; ω is the frequency of rotation of oscillation generator shaft; dW is the volume of fluid supplied to HPH or withdrawn from HPH; *I, II* are the lines for supplying working fluid to HPH

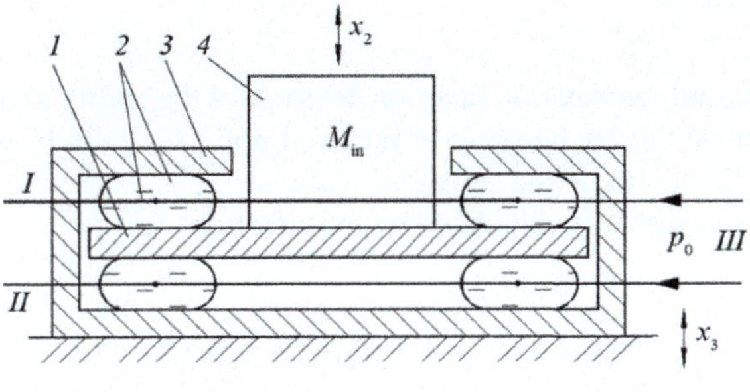

Fig. 6 Actuator device with parallel switching HPH, where 1 is the intermediate plate; 2 is the HPH; 3 is the body; 4 is the inertial mass of the clamp; x_2 is the current coordinate of the transferring the inertial mass; x_3 is the current coordinate of the load on the ground; p_0 is the average pressure of the operating fluid in the system; *I, II* are the lines for supplying the operating fluid to the HPH from the generator; *III* is the line for the formation of average pressure p_0

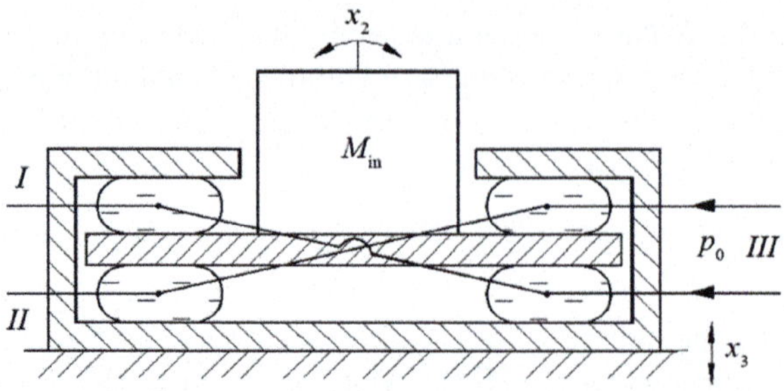

Fig. 7 Actuator device with cross switching HPH, where 1 is the intermediate plate; 2 is the HPH; 3 is the body; 4 is the inertial mass of the clamp; x_2 is the current coordinate of the transferring the inertial mass; x_3 is the current coordinate of the load on the ground; p_0 is the average pressure of the operating fluid in the system; *I*, *II* are the lines for supplying the operating fluid to the HPH from the generator; *III* is the line for the formation of average pressure p_0

- inertial mass, the weight of which is determined by the value of the load transferred to the ground.

The main disadvantage of this actuator is the large weight of load 1 (Fig. 5), which is commensurate in magnitude with the force generated on the ground and determines the weight of the entire source of seismic signals.

A different version of the HPH switching makes it possible to generate a vibration load of various shapes.

3 The Analysis of the Influence of the Amplitude Modulation Law on the Autocorrelation Function Compactness

The degree of the autocorrelation function determines the ability to extract the useful information against the background of received noise on the vibrogram. The more compact the ACF, the better the signal properties.

The ACF of the signal is calculated by the equation

$$(t) \approx \frac{1}{T}\int_{t1}^{t2} F(t)F(t-\tau)dt, \tag{2}$$

where F is the signal transmitted to the ground, T is the signal period, τ is the signal correlation time shift.

At present, sweep signals [30–45] described above are widespread (Fig. 3a).

The purpose of the research is to find out the nature of the influence of the amplitude modulation law on the ACF compactness.

With the same signal parameters (the time-bandwidth product TBP = 14 and duration $T = 0.1$ s) there were studied the signals with different amplitude modulation law:

- "triangular" with increasing amplitude;
- "triangular" with decreasing amplitude;
- "triangular" symmetrical;
- trapezoidal;
- sinusoidal.

According to the results of mathematical modeling, the dependence of the ratio of the decay time of the ACF (Fig. 8) to the time of the signal on the law of amplitude modulation (Fig. 9) was plotted.

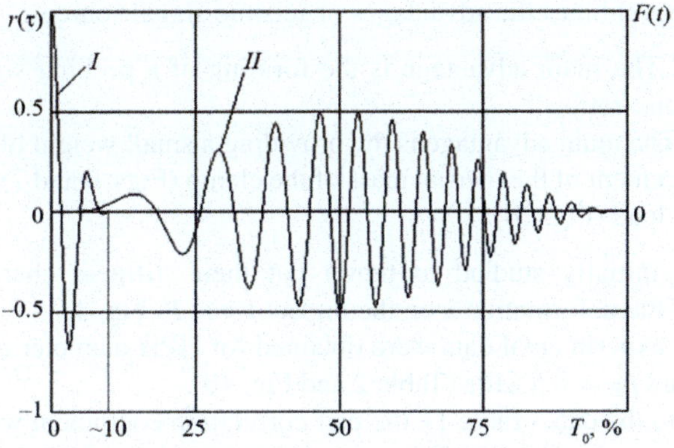

Fig. 8 Results of mathematical modeling, where *I* is the ACF, *II* is the sweep signal with amplitude modulation according to the sinusoidal law

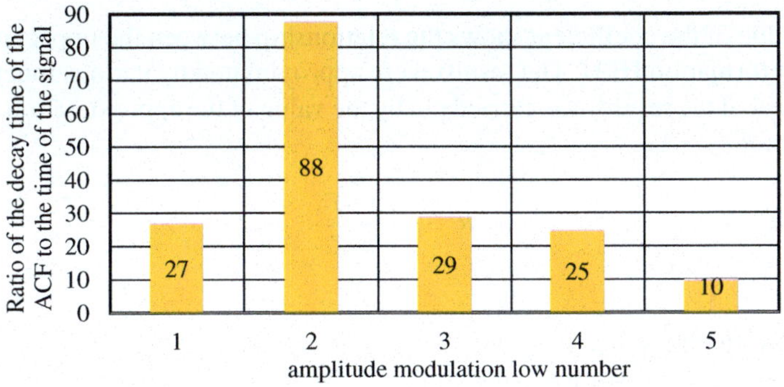

Fig. 9 The dependence of the ratio of the decay time of the ACF to the time of the signal on the law of amplitude modulation: 1 is the "triangular" with increasing amplitude; 2 is the "triangular" with decreasing amplitude; 3 is the "triangular" symmetrical; 4 is the trapezoid; 5 is the sinusoidal

Conclusion: the amplitude modulation of the sweep signal with a sinusoidal law (Fig. 9) makes it possible to obtain a compact ACF. It will provide a better degree of the extraction of useful information against the background of noise on the vibrogram.

4 Shock-Vibration Source of Seismic Signals for the Forming of an Amplitude-Modulated Sweep Signal

The scientists of the National Research Tomsk Polytechnic University proposed a scheme for constructing a shock-vibration source of seismic signals with the ability of generating an amplitude-modulated sweep signal.

This source combines the advantages of seismic signal sources:

- vibrational. The main advantage is the forming of a probing signal with good signal-to-noise ratio;
- impulsive. The main advantage is the providing a small weight of the source due to the replacement of the inertial mass of the clamp (Figs. 6 and 7) with the clamp by a falling load (Fig. 2b).

HPH were initially studied to figure out their stiffness characteristics (the dependence of the deformation x on the impact force F, Fig. 4).

As a result, experimental data were obtained for HPH diameter $d = 22$ mm and average pressure $p_0 = 0.5$ MPa (Table 2 and Fig. 10).

According to the data (Table 1), the pair correlation coefficient was calculated

$$r = \frac{\sum_{i=1}^{7}\left[\left(F_i - \overline{F}\right)\cdot\left(x_i - \overline{x}\right)\right]}{\sqrt{\sum_{i=1}^{7}\left(F_i - \overline{F}\right)^2 \cdot \sum_{i=1}^{7}\left(x_i - \overline{x}\right)^2}} \approx 0.9.$$

The value of the coefficient shows the relationship between the impact magnitude and the deformation HPH. The results were approximated by various functions. The comparison of the results was carried out by the value of the approximation reliability coefficient R^2:

$$R^2 = 1 + \frac{\Sigma_1}{\Sigma_2}, \quad (3)$$

Table 2 Experimental data

Measurement number physical quantity	1	2	3	4	5	6	7
F, kN	−2	−1.5	−0.8	0	1.5	4	10
x	−1.5	−1	−0.5	0	0.5	1	1.5

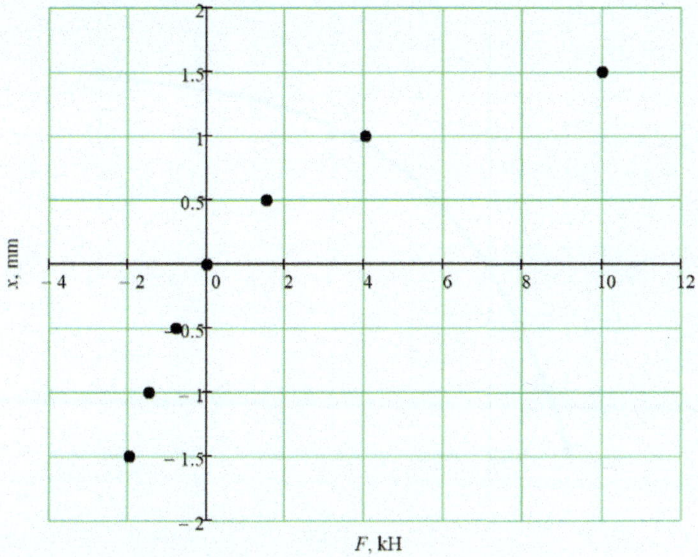

Fig. 10 HPH rigidity characteristic ($d = 22$ мм и $p_0 = 0.5$ MPa)

where $\Sigma_1 = \sum_{i=1}^{n} (x_i - x(F_i))^2$ и $\Sigma_2 = \sum_{i=1}^{n} x_i^2 - \frac{1}{n}\left(\sum_{i=1}^{n} x_i\right)^2$; x_i is the strain values HPH; $x(F_i)$ is the calculated strain values according to the accepted dependence; n is the sample size ($n = 7$).

Good results were obtained with the approximation by the exponential function (Fig. 11).

$$x(F) = a_0 \cdot e^{a_1 \cdot F} + a_2,$$

where $x(F)$ is the approximating function, a_0, a_1, a_2 are the coefficients of the approximating function.

The value of the approximation reliability coefficient was $R^2 = 0.998$. For example, with a parabolic approximation, the value of the approximation reliability coefficient (3) is following:

$$R^2 = 0.979.$$

According to the analysis of the influence of the amplitude modulation law on the compactness of the ACF, one of the possible schemes is proposed for the implementation of a vibrational probing signal and the construction of a vibration source (Fig. 12). The dynamic clamping in the vibration source is carried out due to a force impulse arising from the interaction of a falling load 2 with a shock absorber 7 of the actuator. Under certain assumptions, the force impulse can be taken as a half-wave of a harmonic function (Fig. 13).

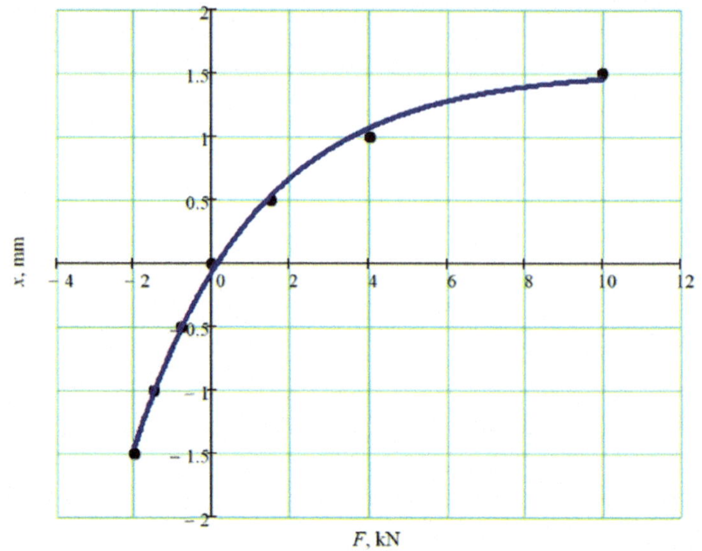

Fig. 11 The results of experimental data approximation by an exponential function

Fig. 12 The scheme of constructing a source of seismic signals: 1 is the housing; 2 is the load; 3 is the cable; 4 is the moving mass; 5 is the limiters; 6 is the HPH; 7 is the shock absorber; 8 is the line; 9 is the hydropneumatic accumulator; 10 is the stops; 11 is the control system; 12 is the controlled hydrostatic oscillation generator; 13 is the line; 14 is the power supply; 15 is the radiating plate

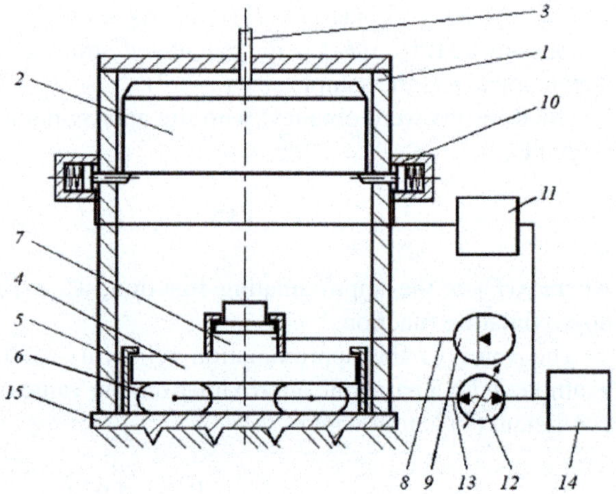

Before starting work, the housing 1 with the emitting plate 15 is installed on the ground at the excitation point. Then the load 2 is lifted to the initial position by the cable 3. The load is fixed with stops 10. The power source 14 is turned on. It drives the adjustable oscillation generator 12. The oscillation generator excites the high-frequency oscillations in the tubular elements 6, which have a non-linear stiffness characteristic (Fig. 11). After a signal from the control system 11 is sent to the stops 10, the load 2 is dropped.

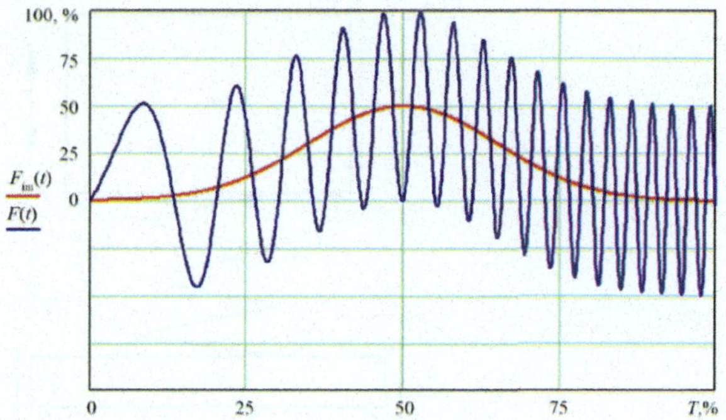

Fig. 13 The law of change of the momentum of the force and the vibrational component

When the load contacts with the shock absorber, a force impulse $F_{\text{im}}(t)$ is excited. The force impulse $F_{\text{im}}(t)$ forms the pressing of the source to the ground (Fig. 12). Before the force impulse $F_{\text{im}}(t)$ from the falling load 2 appears, the force $F(t)$ from the movable mass *4* is transmitted to the ground (Fig. 3a). The movable mass 4 is set to HPH. Vibration action (sweep signal) is excited by oscillation generator 12. During the action of the force impulse, a combined load on the ground $F(t)$ is formed (Fig. 13). The combined load on the ground $F(t)$, depending on the stiffness of the shells (depending on the average pressure p_0), can have a different law changes.

The return of the load 2 to its initial position is carried out by the shock absorber 7 with the energy of the hydropneumatic accumulator 9. The cable 3 is necessary to fix the load 2 only at the beginning of the operation of the device. Thus, the proposed device makes it possible to increase the dynamic impact on the ground, which increases the range of the signal in the ground.

The weight replacing in vibratory mechanisms which is carried out by inertial mass, with a dynamic clamp significantly reduces the weight of the entire machine. It makes possible to obtain a load with an oscillation amplitude of a shape close to sinusoidal. The appliance of a shock absorber with a hydropneumatic accumulator reduces the energy required to return the load to its original position and speeds up the process of generating the next impulse.

Experiments carried out on a laboratory bench (Fig. 14) at an average pressure of $p_0 = 0.5$ MPa showed an increase in the amplitude of the force of the high-frequency component of the signal with each new oscillation period.

For further research, let's make a mathematical model of a real physical system and idealize the properties of the described model. In the future, comparing the obtained modelling results with the experimental results, we will conclude that the applied idealization is acceptable and admissible.

The high-frequency component of the signal reaches its maximum value in the middle of the time interval of the impact pulse F_{im} (Fig. 15).

The studies were carried out according to the mathematical model (not considered in this article).

A fragment of the results is shown (Figs. 16 and 17).

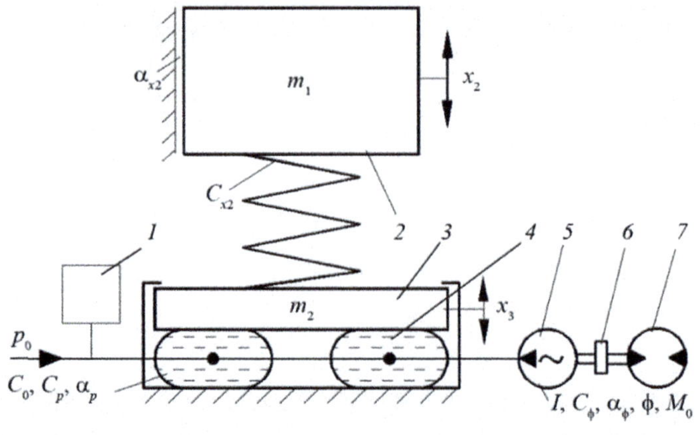

Fig. 14 Schematic diagram of the laboratory bench, where 1 is the sensor; 2 is the load; 3 is the moving mass; 4 is the HPH; 5 is the oscillation generator; 6 is the flywheel; 7 is the hydraulic motor; m_1, m_2 are the weight of the load and the moving part of the actuator; x_2, x_3 are current coordinates of load 2 and moving mass *3* in the mode of joint movement with a shock absorber, moving mass; C_{x2} is the shock absorber hydraulic spring stiffness; α_p, α_{x2} are coefficients of viscous losses in the vibration and shock circuits of the mechanism; C_0 is the stiffness of the main elastic links, which includes the stiffness HPH and the stiffness due to the presence of medium pressure in the cavities; C_p is the hydraulic stiffness due to volumetric deformation; *I*, C_φ, α_φ are the inertia moment, stiffness and losses in the oscillation generator drive circuit; φ, M_0 are phase and moment of the oscillation generator

Fig. 15 The law of change in the force impulse and combined load

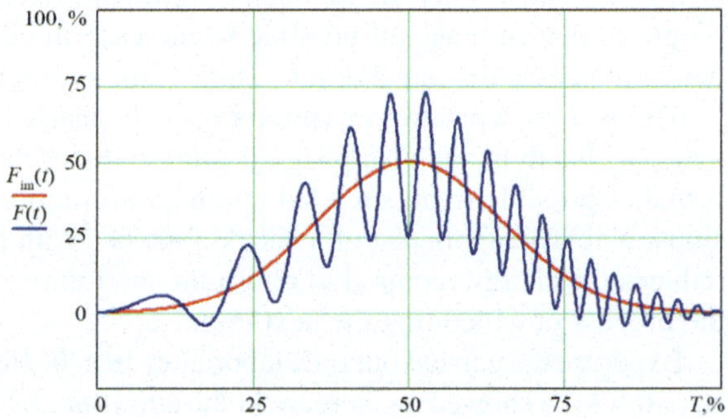

Summarized data on the dependence of the ratio of the ACF decay time to the duration of the signal with amplitude-frequency modulation according to the sinusoidal law δ on its base are shown in Fig. 18.

Pair correlation coefficient $r = -0.749$. The experimental data are well approximated by the dependence ($R^2 = 0.973$):

$$\delta = \frac{a_0}{t} + a_1.$$

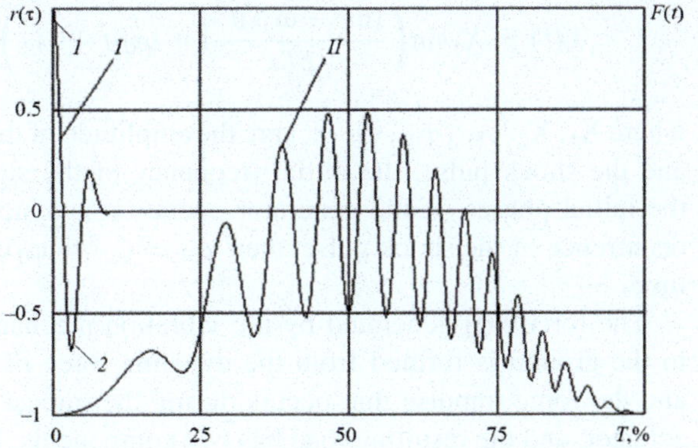

Fig. 16 The law of ACF *I* change and combined load *II* for signal duration $T = 0.1$ s and its base TBP = 14, where 1, 2 are the main and side maxima

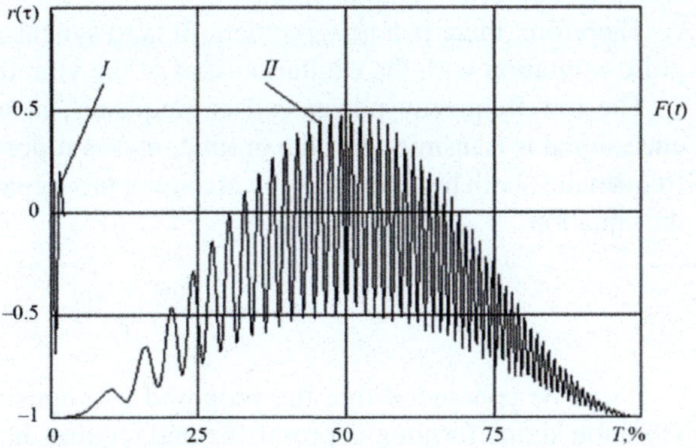

Fig. 17 The law of ACF I change and combined load II at signal duration $T = 0.1$ s and its base TWB = 56

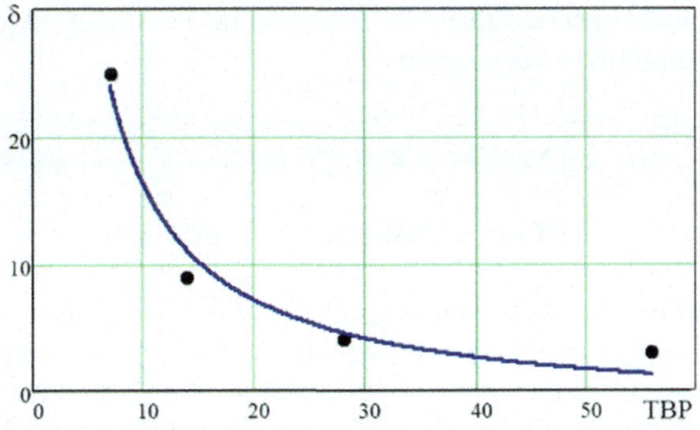

Fig. 18 Modelling data and the results of dependence approximation

Thus, if an ideal probing signal is transmitted to the ground then its ACF (Fig. 16) is compact

$$x(t) = X_1 \sin\left(\frac{(\omega_f - \omega_{01})t^2}{2T} t + \omega_{01} t + \varphi_{01}\right) + X_2 \sin(\omega_2 t + \varphi_{02}); \quad (4)$$

where $X_1, X_2, \omega_{01}, \varphi_{01}, \varphi_{02}, \varepsilon, t$ are the amplitude of the high-frequency sweep signal and the shock pulse, the initial frequency of the signal excited by the generator, the initial phases (if the generator operation is synchronized with the moment of occurrence of the shock pulse, then $\varphi_{01} = \varphi_{02} = \varphi_0$), signal sweep rate and current time.

The force x(t) generated by the vibration mechanism (Fig. 12) and transferred to the ground is formed from the dynamic force of the moving mass 3 vibration and the force impulse that occurs during the impact of the falling load 2 with the actuator, and the resulting load has two components: low-frequency impulse, which provides the dynamic clamping and high-frequency probing sweep signal with good signal-to-noise ratio characteristics.

Therefore, there is a new problem. It is to synchronize the initial moment of the pulse excitation with the excitation start of the vibration signal.

The low-frequency power carrier frequency, at which a high-frequency reference signal is transmitted to the ground, makes it possible to transmit the necessary frequencies, both high and low, by changing the signal sweep rate. It is expressed by the equation

$$\varepsilon = \frac{\omega_f - \omega_{01}}{T}. \quad (5)$$

It can be concluded that the proposed and considered scheme for the probing vibration signal forming is promising and requires more detailed research.

It is impossible in real conditions to generate the exact shape of the sweep signal (1). In this case, it is impossible to emit a signal (4) the required number of times with constant values of signal amplitudes, initial frequency and frequency sweep rate (5) of the sweep signal

$$\begin{aligned} x(t) = &[X_1(t) + \Delta X_1(t)] \cdot \sin\left(\frac{(\varepsilon + \Delta\varepsilon)t^2}{2} + (\omega_{01} + \Delta\omega_{01})t + (\varphi_0 + \Delta\varphi_0)\right) \\ &++[X_2(t) + \Delta X_2(t)] \cdot \sin(\omega_{02} t + \varphi_0) \end{aligned} \quad (6)$$

where X_1, X_2 are the amplitudes of the sweep signal and impulse; ε is the frequency sweep rate of the sweep signal; ω_{01} is the initial sweep frequency of the sweep signal; ω_{02} is frequency of impulse action; ΔX_1, ΔX_2 is the errors in the forming of the amplitude effect, $\Delta\omega_{01}$ is an error in the forming of frequency modulation, $\Delta\varphi_0$ is a synchronization error of the sweep signal and pulse.

Therefore, the task is to access the accuracy degree of generating sweep signal (6) to provide a compact ACF and a clearly defined main maximum.

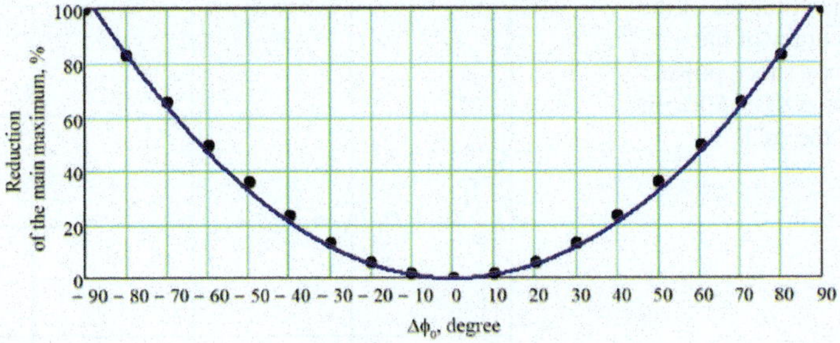

Fig. 19 Modelling data and results of approximation of the decrease dependence in the main maximum against the background of the side one from the phase shift

The computer modelling of the influence of non-synchronous excitation of the sweep signal and the pulse showed a decrease in the main maximum 1 of the correlation function against the background of the secondary maximum 2 with an increase in the phase shift between the shock pulse and the sweep signal (Fig. 16).

According to the results, it was concluded that when the operation is synchronized in the range of $\pm 30°$ the decrease in the main maximum of the correlation function varies in the range of 13% (Fig. 19).

The results of computer modelling can also be approximated with high accuracy ($r = 0.975$, $R^2 = 0.995$) by the Equation

$$0.013 \cdot \Delta\varphi_0^2,$$

taking the value of the decrease in the main maximum in the range $\Delta\varphi_0 > 87.7\%$ and $\Delta\varphi0 < -87.7\%$ as 100%.

The pulse amplitude changing ΔX_2 (Figs. 20 and 21) and sweep rate ε (Figs. 22 and 23) affects the amplitude of side maxima, which can go beyond the exponential curve limiting the correlation function (Fig. 21).

Fig. 20 The influence of the forming accuracy of the low-frequency shock pulse amplitude at $\Delta X_2 = 25\%$ on the type of ACF: *1*, *2* are main and side maxima

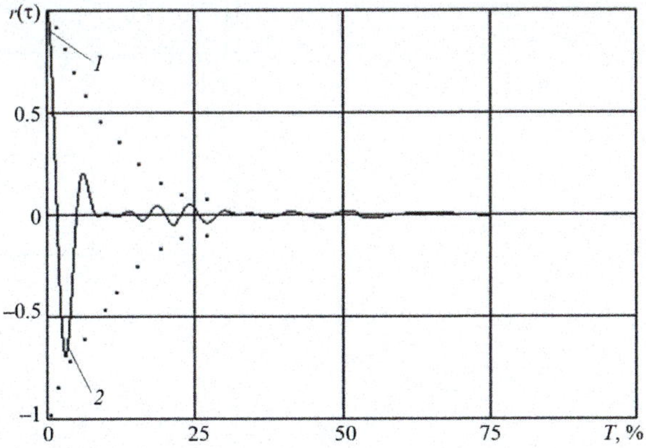

Fig. 21 The influence of the forming accuracy of the low-frequency shock pulse amplitude at $\Delta X_2 = 50\%$ on the form of the ACF

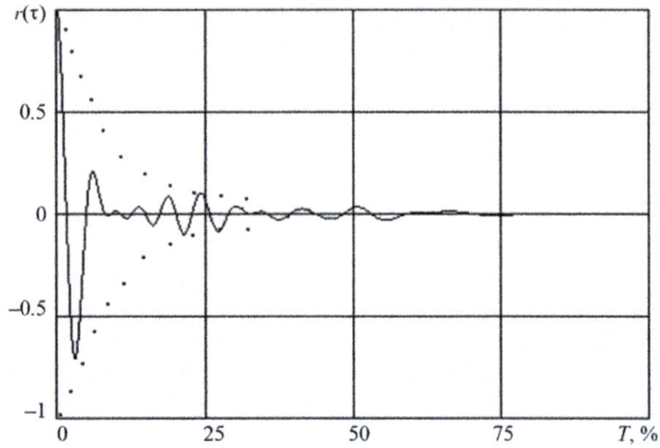

Fig. 22 The influence of the accuracy of the forming of a sweep of a high-frequency sweep signal at $\Delta\varepsilon = \pm 10\%$ on the form of the ACF

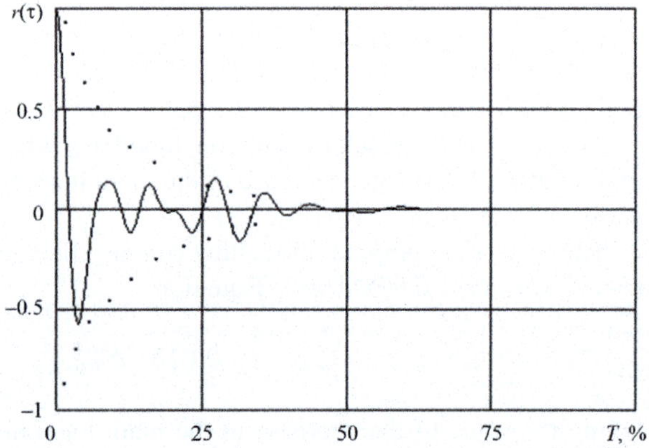

Fig. 23 Influence of the accuracy of the forming of a sweep of a high-frequency sweep signal at $\Delta\varepsilon = \pm 15\%$ on the form of the ACF

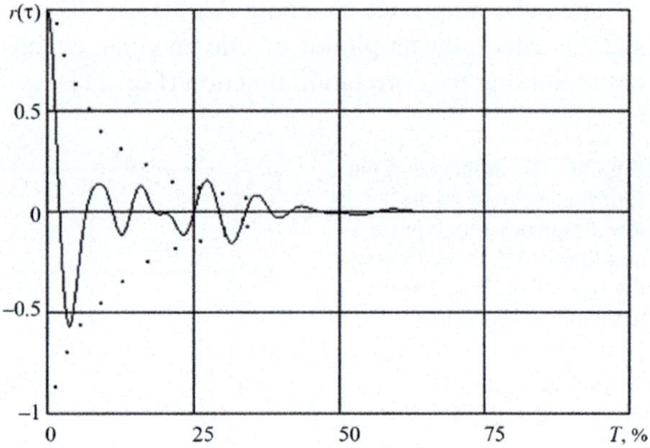

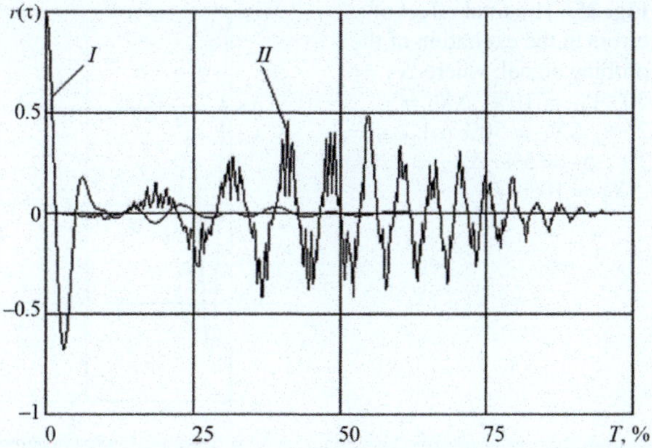

Fig. 24 The influence of the force amplitude of the sweep signal at $\Delta X_1 = 50\%$ on the ACF

An example of the influence of the amplitude of the sweep signal changing on the ACF is shown in Fig. 24.

The initial frequency does not have much effect, because the main parameter of the signal is its base. It remains constant.

The total impact of errors on the quality characteristics of the ACF is shown in Fig. 25.

Thus, if the signal parameters are maintained within the specified limits (Fig. 25b), the signal ACF is relatively compact in the absence of side maxima that go beyond the boundary exponential curve.

For comparison, Fig. 26 shows the total effect of inaccuracies in reproducing a signal with a rectangular amplitude envelope on the qualitative characteristics of the ACF.

Since the influence of inaccuracies during signal excitation is random, it can be assumed that with the accumulation of influences and subsequent correlation, the ACF will meet the requirements for it.

The time-bandwidth product is calculated by the equation

$$TBP = \Delta\omega \cdot T \tag{7}$$

where Δω is a frequency deviation, 140 Hz; T is signal duration, 0.1 s.

With the accumulation of influences (signals) in the amount of n times, the time-bandwidth product is calculated by the equation

$$TBP = \Delta\omega \cdot T \cdot \sqrt{n}. \tag{8}$$

Signals with base (7) were computer studied with the help of computer software

$$TBP = 140 \cdot 0.1 = 14.$$

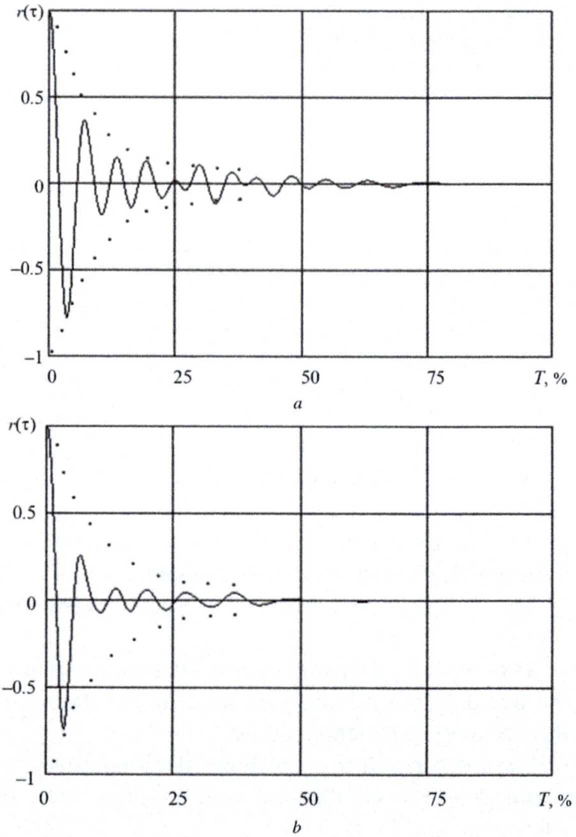

Fig. 25 The total effect of errors in the excitation of the probing signal, where $\Delta\varphi = 30°$, $\Delta\varepsilon = 10\%$, $\Delta X_2 = 25\%$, $\Delta X_1 = 50\%$ (*a*); $\Delta\varphi = 30°$, $\Delta\varepsilon = 5\%$, $\Delta X_2 = 12\%$, $\Delta X_1 = 10\%$ (*b*)

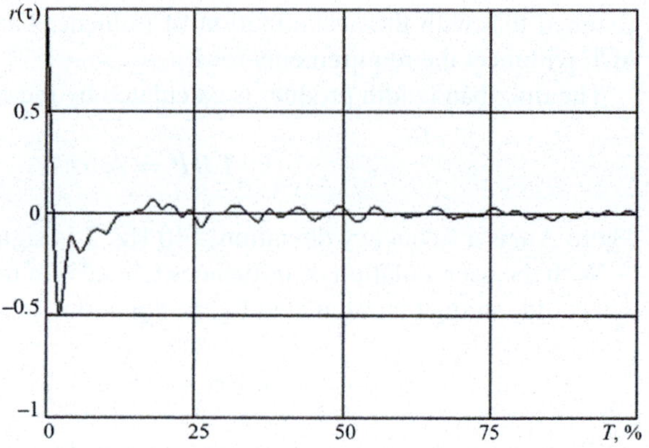

Fig. 26 The total effect of errors in the excitation of a probing signal with a rectangular envelope $\Delta\varphi = 30°$, $\Delta\varepsilon = 5\%$, $\Delta X_2 = 12\%$, $\Delta X_1 = 10\%$

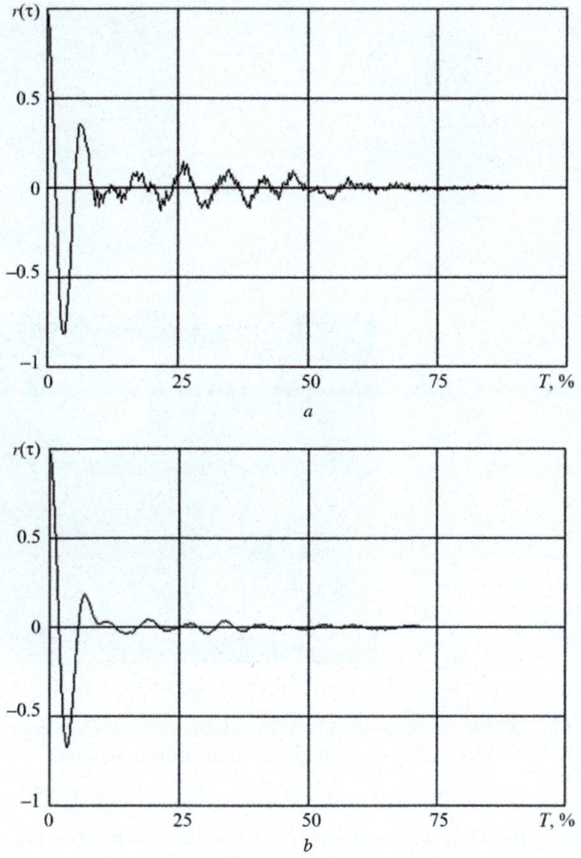

Fig. 27 The total effect of errors in the excitation of a signal with a smooth envelope ($\Delta\varphi = 60°$, $\Delta\varepsilon = 10\%$, $\Delta X_2 = 25\%$, $\Delta X_1 = 50\%$), where TBP = 14 (**a**); TBP = 44.3 (**b**)

Below is an example of the results of a computer study of the ACF of a single signal (Fig. 27a) and after the accumulation of 10 signals (8) (Fig. 27b)

$$TBP = 140 \cdot 0.1 \cdot \sqrt{10} \approx 44.3.$$

5 Experimental Part

According to theoretical information, test and physical experiments were arranged.

Data recording and processing was carried out by a mobile diagnostic complex designed to monitor the current vibration state of technical systems (Fig. 28).

The complex consists of:

- vibration measuring module K-5101 (Ni cDAQ-9174 National Instruments);

a

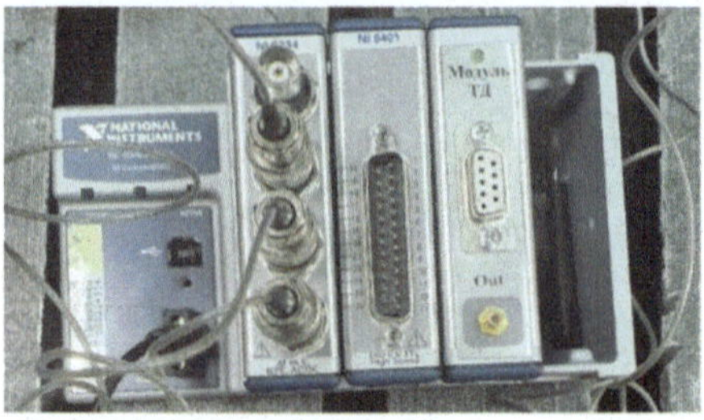

b

c

Fig. 28 Measuring system K-5101: **a** is the vibration converter; **b** is the vibration measuring module K-5101; **c** is the laptop with vibration measuring module

- vibration converters (piezoelectric sensors AR2037-100 "Global Test LLC" Sarov, RF);
- laptop with installed special software "Vibroregistrator-F".

It has been developed the method for selecting the modes of information processing of vibration parameters based on the special software (SpS) "Vibroregistrator-F" to work with the mobile diagnostic complex [46–49].

The method is based on the following principles.

1. Algorithms of multichannel scanning on three coordinates (four points) of the investigated element were used. This provides more informational content when conducting the researches about the law of oscillations.
2. It is possible to adjust the sensors scanning frequency for each channel.
3. The frequency range (2…10,000 Hz) was selected because of carrying out the control in a wide range of vibration parameters: vibration acceleration, vibration velocity, vibration displacement.
4. The dynamic range (at least 110 dB) is resulted from the need to measure the minimum and maximum vibration parameters in the following ranges:

   - accelerations (0.5…500 m/s^2);

- vibration velocities (1…150 mm s);
- vibration displacement (10…1000 μm).

5. To obtain spectrograms in the fast Fourier transform, adjustable number of samples N for averaging at scanning frequencies is used. Also, rectangular window and Hamming window based on experience, due to the need to determine vibration parameters with a relative error in the operating frequency range of no more than 7% and a relative error in the operating amplitude range of no more than 5%.
6. The taken averaging types are:

   - linear (for stationary signals);
   - exponential (for non-stationary signals and transient processes, such as acceleration, coasting).

7. A complex application of adjustable filters (low-pass, high-pass, band-pass, blocking) of Butterward and Chebyshev of the 2nd kind, which have the flattest characteristic in the passband, has been implemented. This allows to minimize the distortion of the analyzed vibration parameters in comparison with other types of filters.
8. It is possible to use vibration converters of various types, depending on the particular range, including non-contact sensors (velometers, proximeters, etc.).
9. The automatic scaling of vibration records and spectrograms was applied (the need for additional adjustment to the actually measured vibration values is excluded).

The research algorithm (Fig. 29) and work with the "Vibroregistrator-F" software are described below.

1. There is setting before measurement:

   - creation of a database (DB) and its addition (Fig. 30);
   - the selection of the name of the device for collecting vibration parameters: cDAQ1Mod1 (Fig. 31a);
   - finding the frequency of the sensor scanning signal: 2048, 5120, 10,240, 25,600, 51,200 Hz (Fig. 31a);
   - the selection of signal update time: 0.2, 0.5, 1 s (Fig. 31a);

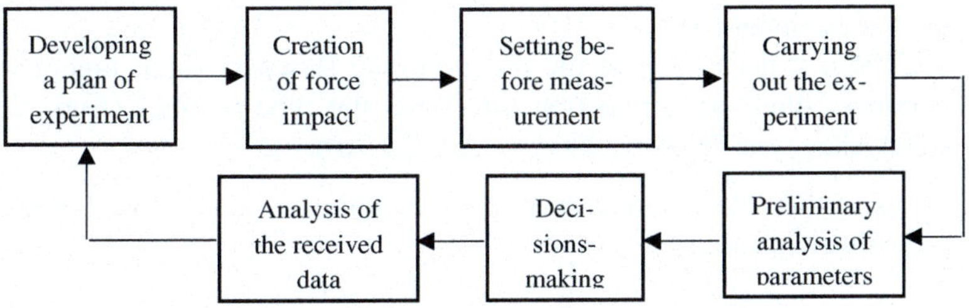

Fig. 29 Research algorithm

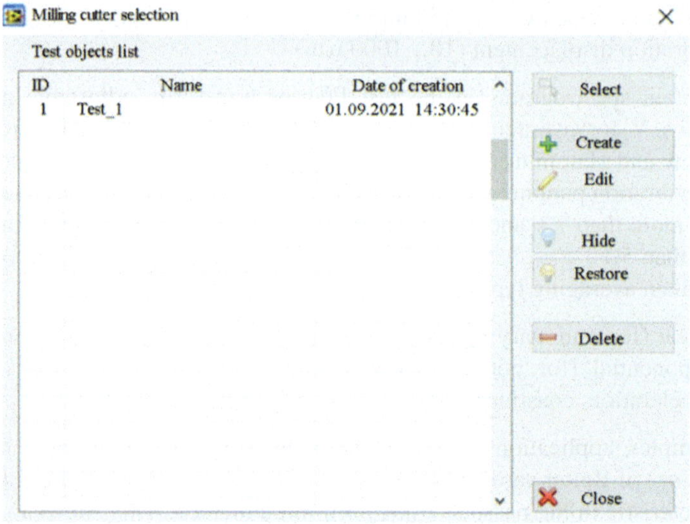

Fig. 30 Database list

- channel selection and color assignment (Fig. 31b)
- the selection of the file save path of test, for example: C\Temp\VITec\ (Fig. 31c);
- assignment of the time interval displayed on the graph (1…100 s), and the maximum recording time (1…0.50 s) (Fig. 31c);
- sensors installation.

2. Carrying out the experiment and recording the information about the tests in the database, carried out according to the developed plan.
3. Extracting the information about tests from the database (Fig. 32) and preliminary analysis conducting.
4. The preliminary analysis of vibration parameters (Fig. 33) includes:

- selection of the time interval: acceleration, operation in a steady state, coasting, as well as the effect of the test influence simulator by a harmonic action or a sweep signal;
- selection of the required vibration parameter: acceleration (m/s^2), speed (mm/s), displacement (μm).
- selection, if necessary, of the required type (lowpass filter, highpass filter, bandpass filter, band-rejection filter) and the type of filter (Butterward or Chebyshev type 2), as well as its order (1…10), etc.

5. Analysis of the data obtained from the creation of a protocol in the Microsoft Word software product.

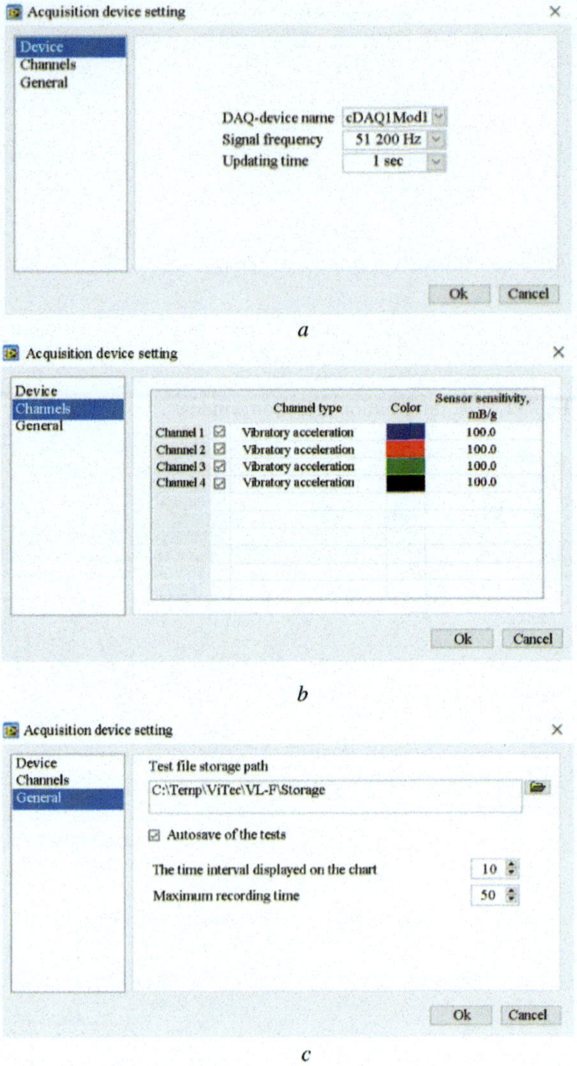

Fig. 31 Acquisition device settings window, where *a* is the tab “Device”; *b* is the “Channels”; *c* is the “General”

The protocol in the form of a report on a single measurement contains information in the form of a table (Table 3) and two diagrams for each information channel (Fig. 34).

From the machining point of view, the analysis of the obtained data can be aimed to determine, for example:

- dependences of the vibration level (peak or RMS) on the modes of machining: cutting speed, tool feed, cutting depth

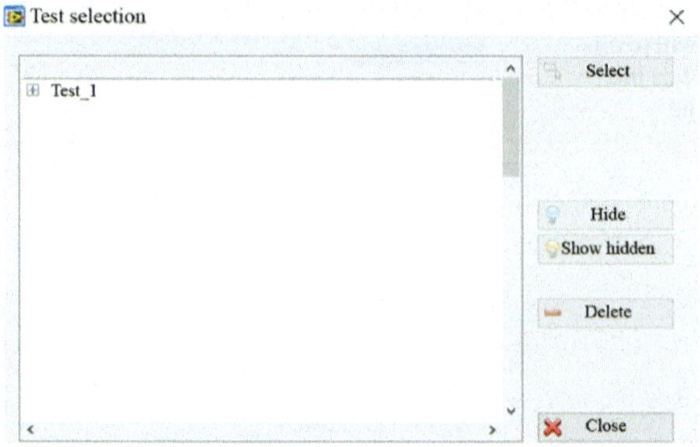

Fig. 32 Test information selection window

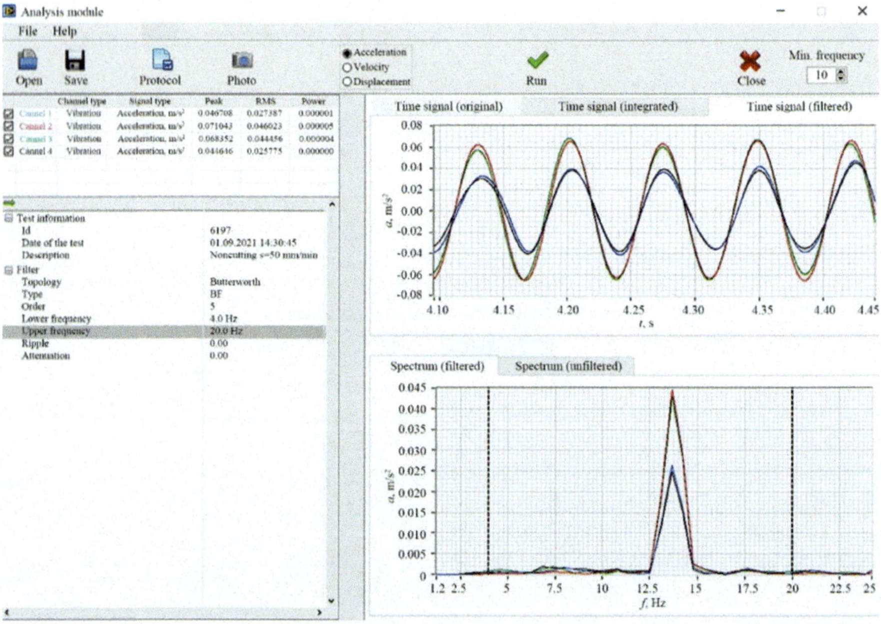

Fig. 33 Analysis module window

Table 3 Data on information channels

Channel	Type	Peak	RMS
1	Acceleration m/s^2	0.046708	0.027387
2	Acceleration m/s^2	0.071043	0.046023
3	Acceleration m/s^2	0.068352	0.044456
4	Acceleration m/s^2	0.044646	0.025775

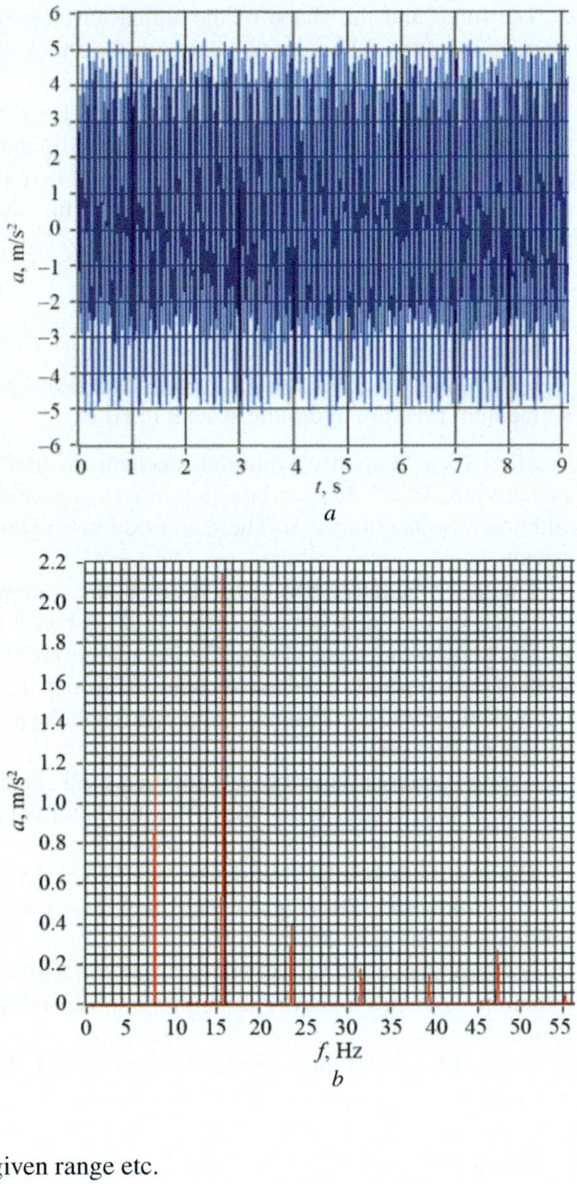

Fig. 34 Experimental diagrams, where **a** is the vibration record; **b** is the spectrogram

- resonant frequencies in a given range etc.

There were considered the probing signals with different shapes of the force amplitude envelope in the previous research section. A signal with a sinusoidal envelope is more promising for application. A constructive scheme of the vibration mechanism including an oscillatory system with a shock assembly was proposed based on this conclusion.

The force and the shape of the signal envelope are formed by the process of interaction of the fallen active load with the shock absorber of the actuator in this scheme of building a vibration source.

The fall of the load was replaced by its rocking-oscillatory motion relative to a certain center in laboratory conditions (to simplify the design). To do that, the load is placed on four cross-connected elastic shells, where the high-frequency oscillations are excited by a symmetrical generator, and the shock mode is simulated by an additional low-frequency generator.

The bench consists of:

- actuator on HPH;
- hydraulic drive of high-frequency generator;
- hydraulic drive of the low-frequency generator;
- medium pressure hydraulic system in HPH.

The hydrovolumetric vibration mechanism, like a number of other hydraulic systems [18, 32, 37–39], can be represented as a generalized dynamic scheme, adding additional connections to it. These connections are due to the presence of a resonant circuit.

The experimental researches of the dynamic system of the mechanism are planned to be carried out on an experimental vibration bench (Figs. 35 and 36).

The bench works as follows. The asynchronous electric motor 1 rotates the input shaft of the adjustable axial-piston pump 3 through the coupling 2. The pump 3 forces the operating fluid (oil) from the tank 5 through the pressure line 4 into the hydraulic system of the stand.

Axial piston hydraulic motor 9 converts fluid energy into mechanical energy. The rotation of the output shaft of the hydraulic motor 9 through the connecting coupling 14 rotates the input shaft of the generator 15.

The generator of high-frequency oscillations 15 supplies the HPH with a variable fluid flow of $\pm dW$ (Fig. 5). This fluid flow generates high-frequency harmonic vibrations of the load 17.

Low-frequency vibrations are generated in a similar way by generator 27. The hydraulic drive has a similar design (not shown in Fig. 36, not yet mounted on the bench).

Figure 37 shows a diagram and the sensors installation photograph.

There was made a plan for conducting the theoretical researches during the process of development and creation of the bench.

The main method of the theoretical study of vibro-impact processes is the exact methods of nonlinear mechanics, but it is a laborious procedure [50] to find the solutions using them. In this regard, the theoretical researches will be carried out on a linear mathematical model of the mechanism.

The comparison of the calculated values of the frequency characteristics with the experimental values will be the method for assessing the validity of the assumptions made for the mathematical model.

A design scheme for a hydrovolumetric vibratory mechanism was developed (Fig. 37) to make a mathematical model.

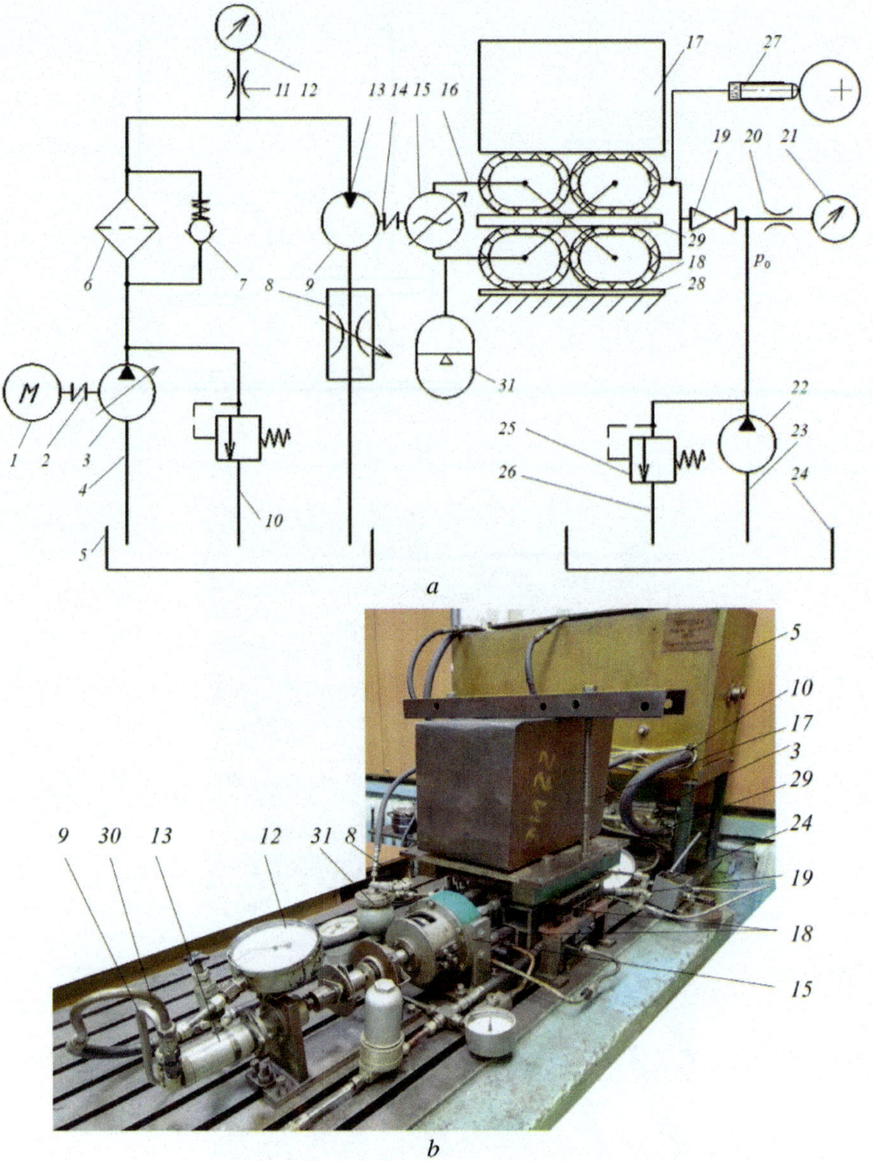

Fig. 35 Experimental bench: **a** is the hydraulic scheme of the experimental bench; **b** is the photo of the experimental bench; 1 is the asynchronous electric motor; 2, 14 are the connecting couplings; 3 is the adjustable axial piston pump; 4, 13, 23 are the pressure lines; 5, 24 are the hydraulic tanks; 6 is the filter; 7 is the check valve; 8, 11, 20 are the throttles; 9 is the axial piston hydraulic motor; 10, 26, 30 are the drain lines; 12, 21 are the manometers; 15 is the adjustable generator of high-frequency oscillations; 16, 18 is the HPH; 17 is the load; 19 is the crane; 22 is the hand pump; 25 is the safety valve; 27 is the adjustable generator of low-frequency oscillations; 28 is the base; 29 is the intermediate plate; 31 is the hydropneumatic accumulator

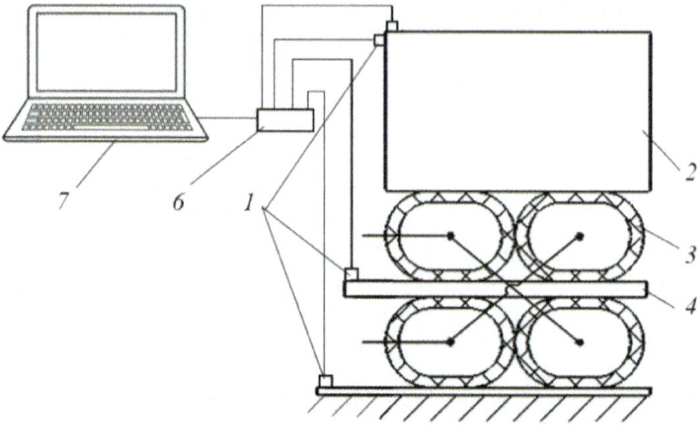

a

b

Fig. 36 Photo of the experimental bench: 1 are the are the vibration converter; 2 is the load; 3 is the intermediate plate; 4 are the HPH; 5 are the piezoelectric sensors; 6 is the vibration measuring module; 7 is the laptop

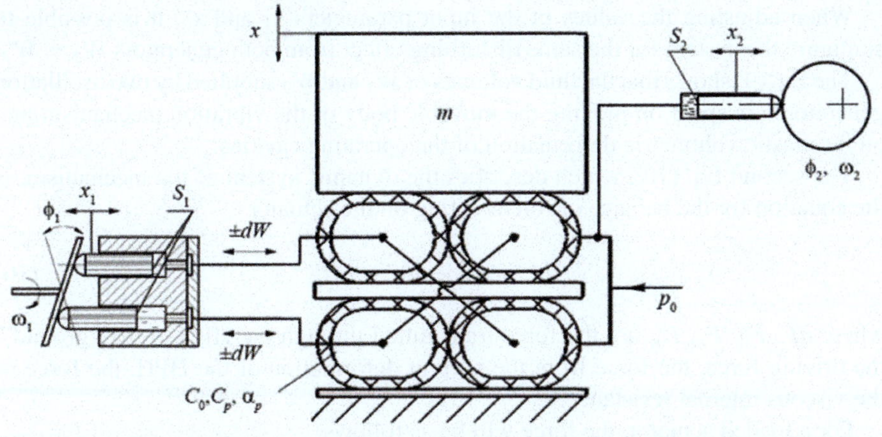

Fig. 37 The design scheme of hydrovolumetric vibromechanism: *m, x* are mass and current coordinate of load moving; S_1, S_2, x_1, x_2 ω_1, ω_2 are: the area of the generators plungers and the coordinates of the plungers moving, the rotational speed of the high-frequency and low-frequency oscillation generators; ω_1, ω_2 are: the area of the generators plungers and the coordinates of the plungers moving, the rotational speed of the high-frequency and low-frequency generators; φ_1 is the tilt angle of the high-frequency generator disk; φ_2 is the current position angle of the low-frequency generator shaft; p_0 is the average pressure in the system; α is coefficient of viscous losses in the vibration circuit of the mechanism; C_0 is the stiffness of the main elastic links, which includes the stiffness HPH C_{hph} and the stiffness due to the presence of medium pressure in the cavities; C_p is the hydraulic stiffness due to volumetric deformation; $\pm dW$ variable fluid flow

When making a mathematical model of a real physical system, we idealize the properties of the described model. Let's make the following assumptions:

- wave processes in hydraulic lines due to their short length do not affect the dynamics of the system;
- there is no undissolved air;
- fluid viscosity coefficient, elasticity modulus, HPH stiffness parameters and operating area of the actuator are constant values;
- there are no liquid leaks from the operating shells and oscillation generators.
- for the unknown variable we will take the coordinate of the "output link" displacement *x*. It characterizes the response of the actuator to the disturbing action of the generators $x_1 = X_1 \cdot \sin\omega_1 t$ или $x_2 = X_2 \cdot \sin\omega_2 t$.

If there are no fluid leaks from the system, the vibration mechanism system is closed. Therefore, the first equation of the mechanism moving is the flow balance equation, which takes into account the property of the continuity of the fluid flow:

$$dW = S \cdot dx = \frac{dp \cdot W_0}{E} = W_1 + W_2 - W; \tag{9}$$

where $W_1 = S_1 \cdot x_1$, $W_2 = S_2 \cdot x_2$ are liquid volumes supplied by high and low frequency generators to the HPH.

When adjusting the values of the input parameters x_1 and x_2, it is possible to configure the system for the same disturbing effect from both generators $W_1 = W_2$.

The Eq. (9) shows that the fluid volumes of W_1 and W_2 supplied by two oscillation generators are spent on moving the movable body of the vibration mechanism and on the elastic volumetric deformation of the operating cavities.

The second Eq. (10), which describes the dynamic system of the mechanism, is the equation for the balance of forces acting on the actuator

$$dF = F_{\mathrm{d}} - F_V - F_x; \tag{10}$$

where dF, F_{d}, F_x, F_V are the force transmitted through the HPH to the "ground", the driving force, the force from the current deformation of the HPH, the force of the viscous internal resistance.

For a loaded actuator, the force will be as follows

$$F_{\mathrm{d}} = C_p \cdot dx = \frac{S^2 \cdot E}{W_0} \cdot dx, \tag{11}$$

where

S	is the operating area of the actuator;
dp	is the pressure drop in operating cavities;
E	is the fluid elasticity modulus;
W_0, dW	is the initial fluid volume in the shells and its change;
dx	is a value of radial deformation of tubular shells.

The force from the current elastic deformation of cavities is as follows

$$F_x = C_0 \cdot x. \tag{12}$$

where

x is the current movement of the moving body;
C_0 is the stiffness of the main elastic bonds

$$C_0 = C_{\mathrm{hph}} + \frac{\pi}{2} \cdot p_0 + \frac{2 \cdot S(x)^2}{W_0 \cdot \beta} = \mathrm{const}. \tag{13}$$

Since the movable body of the vibration mechanism performs a rocking motion, it can be assumed that the magnitude of the amplitudes of its moving in the vertical plane is small, therefore, the change in the area S can be neglected.

The force of viscous internal resistance can be as follows:

$$P_x = \alpha \cdot \frac{dx}{dt}. \tag{14}$$

where α is the coefficient of viscous losses in the oscillatory circuit; V is the movement speed of the movable body of the actuator.

Thus, the general system of equations describing the dynamic system behavior of the vibrator will be as follows:

$$\begin{cases} m\frac{d^2x}{dt^2} + \alpha \cdot \frac{dx}{dt} + C_0 \cdot x = dp \cdot S \\ \frac{E}{W_0}(W_1 + W_2 - W) = dp \end{cases}. \quad (15)$$

Solving the system of Eq. (15), we obtain the general equation of moving (16):

$$m\frac{d^2x}{dt^2} + \alpha \cdot \frac{dx}{dt} + C_0 \cdot x = \frac{E \cdot S}{W_0}(W_1 + W_2 - W);$$

$$m\frac{d^2x}{dt^2} + \alpha \cdot \frac{dx}{dt} + C_0 \cdot x = \frac{E \cdot S \cdot S_1}{W_0} \cdot x_1 + \frac{E \cdot S \cdot S_2}{W_0} \cdot x_2 - \frac{E \cdot S^2}{W_0} \cdot x;$$

$$m\frac{d^2x}{dt^2} + \alpha \cdot \frac{dx}{dt} + C_0 \cdot x = C_p \cdot \frac{S_1}{S} \cdot x_1 + C_p \cdot \frac{S_2}{S} \cdot x_2 - C_p \cdot x.$$

Taking into account $x_1 = X_1 \cdot \sin\omega_1 t$ и $x_2 = X_2 \cdot \sin\omega_2 t$, and by introducing the area ratio designations ($k_{s1} = S_1/S$ и $k_{s2} = S_2/S$) as area gains coefficients, we get as follows:

$$\begin{aligned} & m\frac{d^2x}{dt^2} + \alpha \cdot \frac{dx}{dt} + (C_0 + C_p) \cdot x = \\ & = C_p \cdot k_{s1} \cdot X_1 \cdot \sin\omega_1 t + C_p \cdot k_{s2} \cdot X_2 \cdot \sin\omega_2 t. \end{aligned} \quad (16)$$

6 Conclusion

It was analyzed the influence of the shape of the amplitude envelope of the sweep-signal force on the ACF. It is noted that signals with a sinusoidal envelope have the most compact ACF. The most compact ACF provides the extraction the useful information against the background of received noise. The compactness of the ACF increases with an increase in the time-bandwidth product.

It is considered a new layout of the vibration mechanism with an active load. This load forms a dynamic pressure of the mechanism plate at the moment of excitation of the sweep signal in HPH. In this case, the probing signal is formed from a low-frequency power pulse and a high-frequency sweep component. It is considered the influence of the accuracy of the formation of the sweep signal on the properties of the ACF. There are determined the boundary values of errors of signal parameters, at which its ACF meet the requirement. A structural diagram of the vibration bench has been developed. It is obtained the general equation of the load moving.

References

1. Yao, Z., Luo, M., Mei, J., Zhang, D.: Position dependent vibration evaluation in milling of thin-walled part based on single-point monitoring. Measurement **171**, 108810 (2021)
2. Sinitsa, M.O., Komshin, A.S.: Development of a control system for large-sized products to improve product quality in mechanical engineering. IOP Conf. Ser. Mater. Sci. Eng. **971**, 022057 (2020). https://doi.org/10.1088/1757-899X/971/2/022057
3. Alifov, A.A.: On the calculation by the method of direct linearization of mixed oscillations in a system with limited power-supply. In: Advances in Computer Science for Engineering and Education II, pp. 23–31. Springer Nature Switzerland AG, Cham (2020). https://doi.org/10.1007/978-3-030-16621-2_3
4. Alifov, A.A.: Oscillations for delayed elastic constraint and a limited power energy source. J. Mach. Manuf. Reliabil. **49**(2), 105–109 (2020)
5. Borovskii, G.V., Shavva, M.A., Zakharevich, E.M., Grubyi, S.V., Maslov, A.R.: Ultraprecision machining of brittle optical materials. Russ. Eng. Res. **35**(12), 883–886 (2015)
6. Metelkina, E., Komshin, A.: The use of measuring phase-chronometric systems in the production of cyclic aggregates of aircraft. In: MATEC Web of Conferences, vol. 129, p. 01048 (2017). https://doi.org/10.1051/matecconf/201712901048
7. Irino, N.: Vibration analysis and cutting simulation of structural nonlinearity for machine tool. CIRP Ann. Manuf. Technol. **70**, 317–320 (2021)
8. Grubyi, S.V.: Cutting characteristics in the end milling of stainless steel. Russ. Eng. Res. **37**(2), 125–133 (2017). https://doi.org/10.3103/S1068798X17020083
9. Fan, W., Qiuming, G., Zhigang, L., Haifeng, Q., Cheng, J., Liu, H., Lijun, Y.: Development and application of cutterhead vibration monitoring system for TBM tunneling. Int. J. Rock Mech. Min. Sci. **146**, 104887 (2021)
10. Ahirrao, N.S., Bhosle, S.P., Nehete, D.V.: Dynamics and vibration measurements in engines. Procedia Manuf. **20**, 434–439 (2018). https://doi.org/10.1016/j.promfg.2018.02.063
11. Chávez-García, H., Castillo-Villar, K.K.: Simulation-based model for the optimization of machining parameters in a metal-cutting operation. Simul. Model. Pract. Theory **84**, 204–221 (2018). https://doi.org/10.1016/j.simpat.2018.02.008
12. Feng, J., Wan, M., Gao, T., Zhang, W.: Mechanism of process damping in milling of thin-walled workpiece. Int. J. Mach. Tools Manuf. **134**, 1–19 (2018). https://doi.org/10.1016/j.ijmachtools.2018.06.001
13. Ahmad, M.S., Kukrer, O., Hocanin, A.: Recursive inverse adaptive filtering algorithm. Digit. Signal Process.: Rev. J. **21**(4), 491–496 (2011). https://doi.org/10.1016/j.dsp.2011.03.001
14. Peng, Z.K., Tse, P.W., Chu, F.L.: An improved hilbert-huang transform and its application in vibration signal analysis. J. Sound Vib. **286**(1–2), 187–205 (2005). https://doi.org/10.1016/j.jsv.2004.10.005
15. Seimert, M., Gühmann, C.: Vibration based diagnostic of cracks in hybrid ball bearings. Meas.: J. Int. Meas. Confed. **108**, 201–206 (2017). https://doi.org/10.1016/j.measurement.2017.03.001
16. Urbikain, G., de Lacalle, L.N.L.: Modelling of surface roughness in inclined milling operations with circle-segment end mills. Simul. Model. Pract. Theory **84**, 161–176 (2018). https://doi.org/10.1016/j.simpat.2018.02.003
17. Vogl, G.W., Donmez, M.A., Archenti, A.: Diagnostics for geometric performance of machine tool linear axes. CIRP Ann. Manuf. Technol. **65**(1), 377–380 (2016). https://doi.org/10.1016/j.cirp.2016.04.117
18. Zhang, S., Yu, J., To, S., Xiong, Z.: A theoretical and experimental study of spindle imbalance induced forced vibration and its effect on surface generation in diamond turning. Int. J. Mach. Tools Manuf. **133**, 61–71 (2018). https://doi.org/10.1016/j.ijmachtools.2018.06.002
19. de Souza, A.F., et al.: New mechanistic model to predict machining time for milling free form geometries using 4-axis milling. Procedia CIRP **101**, 34–37 (2021)
20. Li, H., Deng, Z., Golilarz, N.A., Soares, C.G.: Reliability analysis of the main drive system of a CNC machine tool including early failures. Reliab. Eng. Syst. Saf. **215**, 107346 (2021)

21. Sato, R., Shirase, K.: Analytical time constant design for jerk-limited acceleration profiles to minimize residual vibration after positioning operation in NC machine tools. Precis. Eng. **71**, 47–56 (2021)
22. Zhishuang, Z., Jiangchuan, N., Yongjun, S., Shaopu, Y.: Forced vibration of two-degrees-of-freedom machine tool feed system with clearance and friction. Appl. Math. Model. **92**, 281–296 (2021)
23. Gomes, M.C., Brito, L.C., da Silva, M.B., Duarte, M.A.V.: Tool wear monitoring in micromilling using support vector machine with vibration and sound sensors. Precis. Eng. **67**, 137–151 (2021)
24. Lysenko, E.N., Nikolaev, E.V., Vlasov, V.A., Surzhikov, A.P.: Microstructure and reactivity of Fe_2O_3–Li_2CO_3–ZnO ferrite system ball-milled in a planetary mill. Thermochim. Acta **664**, 100–107 (2018). https://doi.org/10.1016/j.tca.2018.04.015
25. Surzhikov, A.P., Lysenko, E.N., Vasendina, E.A., Sokolovskii, A.N., Vlasov, V.A., Pritulov, A.M.: Thermogravimetric investigation of the effect of annealing conditions on the soft ferrite phase homogeneity. J. Therm. Anal. Calorim. **104**(2), 613–617 (2011). https://doi.org/10.1007/s10973-010-1074-4
26. Surzhikov, A.P., Frangylyan, T.S., Ghyngazov, S.A.: A dilatometric study of the effect of pressing on the kinetics of compression of ultrafine zirconium doxide powders under thermal annealing. Russ. Phys. J. **55**(4), 345–352 (2012). https://doi.org/10.1007/s11182-012-9818-1
27. Czichos, H.: Handbook of technical diagnostics: Fundamentals and application to structures and systems. In: Handbook of Technical Diagnostics. Springer Heidelberg, New York (2013)
28. Edward Smith, H. (ed.): Mechanical Engineer's Reference Book. Butterworth-Heinemann, Oxford (1994)
29. Lockhart, B.: Mechanical Engineering Reference Manual. NY Research Press, New York (2015)
30. Cao, Y., Osadchiy, A.V., Xin, X., Yin, X., Yu, C., Zhang, X., Monroy, I.T.: Performance analysis of IM, DPSK and DQPSK payload signals with frequency swept coherent detected spectral amplitude code labelling. Opt. Switch. Netw. **8**(2), 79–85 (2011). https://doi.org/10.1016/j.osn.2010.11.003
31. Ogino, T., Mitachi, T., Oikawa, H., Takahashi, T., Tsushima, M.: Frequency response characteristics of bender element test system identified by frequency-swept signal input—Influence on accuracy of received waveform reconstruction. Soils Found. **50**(5), 737–745 (2010). https://doi.org/10.3208/sandf.50.737
32. Ogino, T., Mitachi, T., Chan, K.H., Oikawa, H., Tsushima, M.: A method for received waveform reconstruction based on bender element test using frequency-swept signal. Soils Found. **48**(2), 287–295 (2008). https://doi.org/10.3208/sandf.48.287
33. Li, P., Zhang, Y., Yao, J.: Linear frequency swept laser source with high swept slope based on digital optical phase-locked loop. Optics Commun. **525**, 128860 (2022). https://doi.org/10.1016/j.optcom.2022.128860
34. Tang, X., Suddarth, S., Kantesaria, S., Garwood, M.: A frequency-swept, longitudinal detection EPR system for measuring short electron spin relaxation times at ultra-low fields. J. Magn. Reson. **342**, 107279 (2022). https://doi.org/10.1016/j.jmr.2022.107279
35. Lu, B., Cheng, J., Hu, J., Qin, S.: Seismic features of vibration induced by mining machines and feasibility to be seismic sources. Procedia Earth Planet. Sci. **3**, 76–85 (2011). https://doi.org/10.1016/j.proeps.2011.09.068
36. Wang, L., Li, H.: GBO algorithm for seismic source parameters inversion. Geod. Geodyn. (In Press) (Available online). https://doi.org/10.1016/j.geog.2022.06.004
37. Duan, Y., Luo, X., Si, G., Canbulat, I.: Seismic source location using the shortest path method based on boundary discretisation scheme for microseismic monitoring in underground mines. Int. J. Rock Mech. Min. Sci. **149**, 104982 (2022). https://doi.org/10.1016/j.ijrmms.2021.104982
38. Lindeberg, L., Ljungberg, R.Y., Leighton, W.M.: A high-order finite-difference scheme to model the fluid-structure interaction in pneumatic seismic sources. J. Comput. Phys. **424**, 109849 (2021). https://doi.org/10.1016/j.jcp.2020.109849

39. Sitong, Y., Jiulong, C.: Numerical simulation of fore detecting with seismic in coal roadway and study of wave field characteristics. J. China Coal Soc. **35**(10), 1633–1637 (2010)
40. Poletto, F., Schleifer, A., Zgauc, F., Meneghini, F., Petronio, L.: Acquisition and deconvolution of seismic signals by different methods to perform direct ground-force measurements. J. Appl. Geophys. **135**, 191–203 (2016)
41. Surzhikov, A.P., Galtseva, O.V., Vasendina, E.A., Vlasov, V.A., Nikolaev, E.V.: Processing line for industrial radiation-thermal synthesis of doped lithium ferrite powders. IOP Conf. Ser.: Mater. Sci. Eng. **110**(123), 012002 (2015). https://doi.org/10.1088/1757-899X/110/1/012002
42. Surzhikov, A.P., Lysenko, E.N., Vlasov, V.A., Malyshev, A.V., Nikolaev, E.V.: Investigation of the process of ferrite formation in the Li_2CO_3–ZnO–Fe_2O_3 system under high-energy actions. Russ. Phys. J. **56**(6), 681–685 (2013). https://doi.org/10.1007/s11182-013-0085-6
43. Surzhikov, A.P., Pritulov, A.M., Lysenko, E.N., Sokolovskii, A.N., Vlasov, V.A., Vasendina, E.A.: Dependence of lithium-zinc ferrospinel phase composition on the duration of synthesis in an accelerated electron beam. J. Therm. Anal. Calorim. **110**(2), 733–738 (2012). https://doi.org/10.1007/s10973-011-1947-1
44. Lysenko, E.N., Malyshev, A.V., Vlasov, V.A., Nikolaev, E.V., Surzhikov, A.P.: Microstructure and thermal analysis of lithium ferrite pre-milled in a high-energy ball mill. J. Therm. Anal. Calorim. **134**(1), 127–133 (2018). https://doi.org/10.1007/s10973-018-7549-4
45. Richter, H., Hock, S., Mikulla, S., Krüger, K., Lüth, S., Polom, U., Dickmann, T., Giese, R.: Comparison of pneumatic impact and magnetostrictive vibrator sources for near surface seismic imaging in geotechnical environments. J. Appl. Geophys. **159**, 173–185 (2018). https://doi.org/10.1016/j.jappgeo.2018.08.010
46. Information. http://www.labview.ru/solutions/172/item3228
47. Gavrilin, A.N., Moyzes, B.B., Kuvshinov, K.A., Vedyashkin, M.V., Surzhikova, O.A.: Determination of optimal milling modes by means of shock-vibration load reduction on tool and peak-factor equipment. Mater. Sci. Forum **942**, 87–96 (2019). https://doi.org/10.4028/www.scientific.net/MSF.942.87
48. Nizhegorodov, A.I., Gavrilin, A.N., Moyzes, B.B., Cherkasov, A.I., Zharkevich, O.M., Zhetessova, G.S., Savelyeva N.A.: Radial-piston pump for drive of test machines. IOP Conf. Ser.: Mater. Sci. Eng. **289**(1), 012014 (2018). https://doi.org/10.1088/1757-899X/289/1/012014
49. Nizhegorodov, A.I., Gavrilin, A.N., Moyzes, B.B.: Application and production technology of thermal activation products of serpentine minerals from industrial wastes Bulletin of the Tomsk Polytechnic University. Geo Assets Eng. **329**(5), 67–75 (2018)
50. Harris, C., Piersol, A.: Harris' Shock and Vibration Handbook. McGraw-Hill Professional, New York (2002)

Usage Practice of Information Technology for the Reorganization of Production Processes

Irina Kolchurina, Maria Kolchurina, Renat Khamitov, and Inna Plotnikova

Abstract One of the trends of modern economics is informatization is the use of information technologies for the collection, storage, processing of data for further use in professional and other types of human activity. The paper shows the development of an information system with the formulation of the customer's primary requirements for the information system and the assessment of existing analogues. The implementation of the documented procedure is described in detail. The designed information system is intended for use by users who are directly involved in production processes and the implementation of corrective actions. Measurements of the time of the flow of the stages of the process were carried out for five cycles of the production process, the average time of the process was determined. It has been proved that the informatization of production processes allows reducing the time for auxiliary processes, thereby increasing the efficiency of the main production processes directly related to the manufacturing technology of products.

1 Introduction

In modern times, enterprises of various sizes pay considerable attention to information technology to maintain the market competitiveness of their own products or services, reduce the cost of production, and reduce reputational risks because of reducing the cost of non-compliance and more appropriate allocation of costs for compliance. Reorganization of production processes allows solving the problems of

I. Kolchurina · M. Kolchurina
Department of Quality Management and Innovation, Siberian State Industrial University, Novokuznetsk 654000, Russia

R. Khamitov
Department of Computer Science and Information Management Systems, Kazan State Power Engineering University, Kazan 420000, Russia

I. Plotnikova (✉)
School of Non-Destructive Testing, Tomsk Polytechnic University, 30, Lenin Avenue, Tomsk 634050, Russia
e-mail: inna@tpu.ru

E. Lysenko et al. (eds.), *Emerging Trends in Materials Research and Manufacturing Processes*, Engineering Materials, https://doi.org/10.1007/978-3-031-38964-1_8

inefficiency in the use of any processes in a timely manner [1]. The use of information technologies underlies the development of Industry 4.0, the deployment of which began in 2013 and continues to nowadays [2, 3].

Large companies use cloud services that make it possible to calculate big data and simplify remote experimental work [4]. For example, Jira and Slack are among the most functional and popular teamwork services. Jira provides powerful functionality for problem solving, time management, statistics gathering, and reporting management. Slack is focused on working within a team, sharing files, thematic channels.

One of the trends of modern economics is informatization is the use of information technologies for the collection, storage, processing of data for further use in professional and other types of human activity.

The introduction of information technologies in the implementation of various production processes and individual operations allows to reduce losses caused by the human factor, speed up the time of processes execution, reducing the number of processing and waiting losses, increase the efficiency of the organization as a whole and expand the scope of activities by starting work in conditions unsuitable for human functioning [5, 6]. Informatization affects many areas of labor activity, including production, mining operations, including in the conditions of the Far North, as well as the provision of service sector, trade, medical assistance and education [7–9]. The use of new technologies is also one of the methods of transformation during the implementation of the lean production concept, which involves reducing the time and the total number of actions that are not aimed at creating value for which the consumer pays, including waiting time, the number of movements of employees of the organization and other aspects [10].

One of the main directions of the introduction of information technologies is the modernization of information exchange procedures between the executors of business processes, including document management procedures [11]. This topic became particularly relevant in the era of the COVID-19 pandemic, when many of the employees refused to work at their workplace, changed the format of work to remote [12]. To ensure the stable functioning of the organization, management had to look for and implement new work technologies that provide sufficiently fast online communications [13].

This problem also affected aspects of the management system process management, one of which is the evaluation of the management system through the use of the audit method.

2 Research Method

During the study the internal audit process was considered, one of the mandatory processes in organizations that have implemented and maintain a quality management system according to the ISO quality management system model. Audit is a systematic, independent and documented process for obtaining audit evidence and evaluating it objectively to determine the extent to which the audit criteria are fulfilled [14]. A distinctive feature of internal audit is that auditors—persons who conducts an audit [15] are employees of the organization. That is why internal audit often call as first party audits. Results of the studies are based on thematic literature and results of modeling.

The process begins with such an activity as the managing an audit program, which includes such operations as determination the objectives for the audit program, development the audit program, agreement the audit program with further approval. During the management of the audit program, the person responsible for managing the audit program determines the scope of the audit program, the scope and criteria of individual audits of the program, identifies and evaluates the risks of the audit program, determines the need for resources, forms an audit group and approves the head of the audit group. The program is approved by the order of the top management and communicated to the staff of the organization.

After approval the audit program, the audit team prepares for the audit in the departments of the organization. During this part of audit activity, audit team must analyze the management system documentation and collect information for the preparation of audit activities and assessment of compliance of the management system with requirements. At this stage, the audit working documents necessary for recording and documenting the auditor's certificates are also being developed, for example, checklists, sample plans and other documents that are determined by the audit team as necessary.

After the completion of the activities for preparing for the audit team start to work. On-site internal audit activities include holding a preliminary meeting, the work of auditors in the audited units, collecting audit evidence by observing or interviewing employees with further logging, exchanging information during the audit, analyzing audit evidence, forming audit conclusions that contain general characteristics of the management system, evaluating its effectiveness and efficiency, descriptions of detected inconsistencies and recorded notifications. Based on the audit findings, the audit team forms conclusions on the audit results, which are presented by audit team leader at the final meeting. At the final meeting, all disagreements between the audit team and delegates of the audited unit are also resolved.

Based on the audit report, audit team leader prepares and sends out an audit report, which includes data on the objectives and scope of the audit, members of the audit team and representatives of the audited organization, audit dates, audit criteria and conclusions, audit conclusions. If necessary, additional information is included in the report, for example, information about contradictions that have arisen, information about opportunities for improvement. After completing all the activities provided for in the audit program, it is considered completed. The audit team is disbanded, and a set of documents used during the audit or containing data on its results are transferred to the archive for storage.

To eliminate inconsistencies identified during the audit, a representative of the audited unit develops a corrective action plan. which includes the name of the unit being checked, descriptions of the detected inconsistencies, their causes, proposed corrective actions, resources required for their implementation, persons responsible for their execution. The corrective action plan is approved by the top management, a representative of the quality management and responsible for internal audit. After the expiration of the deadlines for the execution of corrective actions fixed in the corrective action plan, a re-check is carried out for the implementation of corrective actions. It is also possible to check the status of corrective actions during the next internal audit. Based on the results of the implementation of corrective actions, their analysis is carried out in order to determine the results of the corrective actions taken. The analysis is carried out by the person responsible for the internal audit and the heads of the relevant structural units involved in the implementation of corrective actions. Based on the results of the analysis, a report is compiled containing information about the results of corrective actions, indicators of their effectiveness.

To visualize the progress of the process and the distribution of roles between the participants was developed "use-case" diagram in the UML language. The UML language includes a set of tools that allow for a comprehensive analysis of complex projects, both from a technical point of view and from the point of view of the needs of business customers [16]. The "use case" diagram describes the interaction between the user and the system by highlighting the use cases that are possible in the described system and the actors—persons interacting with the system. The resulting diagram is shown in Fig. 1.

3 Description of the Installation

The implementation of the internal audit process is associated with such information flows as the transfer of information from the top managment of the organization to the person responsible for internal audit on the timing, scope and criteria of the audit; the transfer of information about the audit and the progress of its implementation from the audit team members to audit group leader; the result of work with inconsistencies from the heads of structural divisions to audit team leader; the collection of information about inconsistencies and audit certificates from members of the audit group to the head of the audit group [17]. In the course of evaluating the effectiveness of the process, the timekeeping method was applied and the average time of the process was determined based on the results of measurements of the time of the stages of the process for five cycles of internal audit. The results are shown in Table 1.

The efficiency of the process was 55%, while the duration of operations not aimed at creating value was 2840 min.

It was noted that a lot of time is spent on the exchange of information between the participants of the process and the coordination of various documentation, also in some cases, the audit team members make mistakes during the filling out of forms, which requires additional time for their processing, so it was decided to automate this process by developing and commissioning an information system.

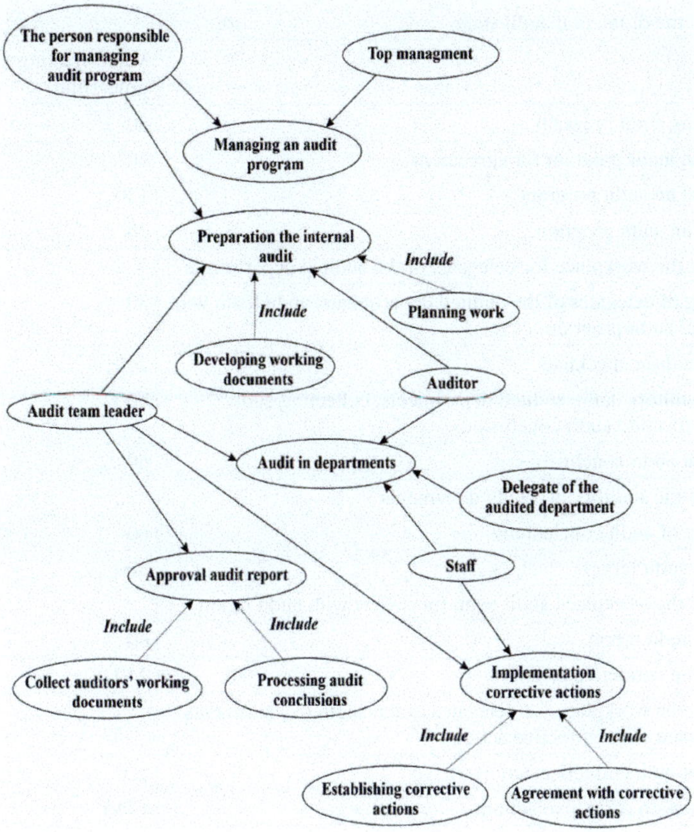

Fig. 1 Process control options

4 Results and Considerations

It was noted that a lot of time is spent on the exchange of information between the participants of the process and the coordination of various documentation, also in some cases, the audit team members make mistakes during the filling out of forms, which requires additional time for their processing, so it was decided to automate this process by developing and commissioning an information system.

The development of the information system began with the formulation of the customer's primary requirements for the information system and the evaluation of the available Russian-language analogues, which in some parameters did not fit the customer's descriptions. At the next stage, the requirements were specified and documented, and later implemented using such programming language as Java and the DBMS PostgreSQL, the choice of which was determined by the customer's requirements and the convenience of work in the implementation of work on the creation of an information system.

Table 1 Time of internal audit stage

Activity	Value creating time, (min)	Muda, (min)
Establishing audit program	30	210
Sending an audit program for agreement	10	30
Agreement an audit program	120	100
Approval an audit program	20	60
Waiting at the workplace for delegates of the audited departments	–	300
Reviewing of delegates of the audited departments, audit team with the internal audit program	30	
Creating auditor checklists	240	80
Work of auditors in the audited departments, collecting audit evidence, forming audit conclusions	600	
Agreement audit conclusions	60	240
Collection the auditors' working documents		240
Processing of audit conclusions	60	600
Preparing audit report	300	
Waiting at the workplace audit team for review with audit report		120
Approval audit report	60	
Establishing corrective actions	240	240
Waiting at the workplace for delegates of the audited departments for agreement with corrective actions		240
Agreement with corrective actions	120	200
Implementation corrective actions	1440	
Evaluation of the effectiveness of corrective actions	120	180

The developed information system has a variety of functions. The list of available functions differs depending on the status of the user: such statuses were allocated as "audit team leader", which could be obtained by the head of the audit team; "auditor", corresponding to a member of the audit team; "administrator", performing system maintenance, "user".

1. The function of creating document forms using Unicode encodings, contain gaps to fill in and a basis that is not editable by users with the status of "auditor". It is possible to create forms of documents such as an audit program, a checklist, and a corrective action plan.
2. The function of entering information using Cyrillic alphabet, numbers from 1 to 5, as well as JPEG format for audit photos.
3. The data storage function in the database. The database stores information about audits of past periods and accounting documents information about auditors (such as name, place of work, qualification, and audit experience), information about static and effectiveness of corrective actions, which were planned after previous audits. Description of database fields presented in Table 2.

Table 2 Description of database fields

Field content	Name of field	Type, length	Constrains
Table audit			
Code of audit	A_ID	N (4)	Primary key
Year	A_YEAR	N (4)	Required field
Audit team	A_PERSON	C (300)	Multivalued field
No. audit program	A_PLAN	N (5)	Multivalued field
No. audit conclusion	A_PROT	N (6)	Multivalued field
Status of corrective actions	A_QM	C (30)	Multivalued field
Table auditor			
Code of auditor	P_ID	N (4)	Primary key
Department	P_PLACE	C (50)	Foreign key
No. certificate of professional development	P_EXP	C (20)	Required field
Experience	P_SKILL	N (2)	
Table department			
Code of department	PL_ID	N (4)	Primary key
Name	PL_NAME	C (50)	Required field
Table type of noncompliance			
No. Type of noncompliance	N_ID	N (4)	Primary key
Name	N_NAME	C (50)	Required field
Table audit conclusion			
No. audit conclusion	U_ID	N (4)	Primary key
Date	U_DATE	D	Required field
Department	U_PLACE	C (50)	Required field
Type of noncompliance	U_TYPE	C (50)	Required field
Description	U_DEC	C (300)	Required field
Table corrective actions			
No. corrective actions	N_ID	N (4)	Primary key
Department	N_NAME	C (50)	Required field
Period	N_DAY	C (30)	Required field
Description	N_MORE	C (600)	Required field
Status of corrective actions	N_S	C (60)	Required field
Table status of corrective actions			
No. status of corrective actions	S_ID	N (4)	Primary key
Name	S_NAME	C (60)	Required field
Table audit criteria			
Code of critiria	K_ID	N (6)	Primary key
Name	K_NAME	C (60)	Required field

(continued)

Table 2 (continued)

Field content	Name of field	Type, length	Constrains
Table audit program			
No. audit program	L_ID	N (6)	Primary key
Date	L_DATE	D	Required field
Author	L_PER	C (50)	
Audited departments	L_PLACE	C (50)	Required field

4. The function of generating a report based on the data entered in the protocols.
5. The function of familiarization with the results of the audit, which is to provide users with the opportunity to familiarize themselves with the protocols of nonconformities, the report on the results of the audit and the corrective action plan.
6. The corrective action management function, which includes the possibility of reviewing and approving the corrective action plan, as well as evaluating the effectiveness of a particular corrective action (as a percentage for users with the status of "auditor" and in the variable "completed/not completed" for "users").
7. Function of information output in.doc format.

Modes of operation of the information system:

- regular mode, in which all functions are available from 9:00 to 20:00 from Monday to Friday, according to the work schedule of the quality management department;
- emergency mode, in which data input and output is unavailable, an administrator call is required to set up the system. It is mandatory to keep logs of fixing accidents in electronic form according to the form established in the organization.

In the normal mode, the information system in one session should provide the following quantitative indicators that characterize the degree of compliance with its purpose:

- the number of simultaneously processed documents—up to 100;
- the number of simultaneous users—up to 20.

Consider the requirements for the types of security. There are no requirements for mathematical and metrological support. When determining the requirements for information support, the requirements for the formats of input information that should be received in Russian were formulated. When filling out checklists, numbers from 1 to 5 should be used. Photos should be uploaded in JPEG format up to 1 MB in size. The interface language is Russian. Encoding of information is not required.

The information system must be able to be installed on Microsoft OS 7 and above; for smartphones and tablets on Android OS 7 and above or iOS 11 and above.

The designed information system is intended for its further use by users who are directly involved in the audit and the implementation of corrective actions. Therefore, for the convenience of using the system in operation, a clear and pleasant interface was developed.

Table 3 Time of internal audit stages after automation

Activity	Value creating time, (min)	Muda, (min)
Establishing audit program	120	
Sending an audit program for agreement	4	
Agreement an audit program	120	
Approval an audit program	20	60
Reviewing of delegates of the audited departments, audit team with the internal audit program	30	
Actualization auditor checklists	240	
Work of auditors in the audited departments, collecting audit evidence, forming audit conclusions	600	
Agreement audit conclusions	60	
Collection the audit working documents	5	240
Processing of audit conclusions	60	
Preparing audit report	120	
Approval audit report	60	
Establishing corrective actions	240	
Waiting at the workplace for delegates of the audited departments for agreement with corrective actions		30
Agreement with corrective actions	120	
Implementation corrective actions	1440	
Evaluation of the effectiveness of corrective actions	120	

According to the results of the implementation of the information system, a decrease in the process time was noted due to the use of an electronic system of coordination and familiarization from 6290 to 3869 min, and the procedure for preparing for external inspections was also optimized. The efficiency of the process after automation was 91%. The time of the individual stages of the process is shown in Table 3.

5 Conclusion

Thus, in the course of the work, the internal audit process in the organization was informatized through the development and implementation of a specialized information system. The implementation of this event allowed to increase the efficiency of the process by 37%, reduce its duration and decrease labor intensity, as well as optimize procedures aimed at preparing for second- and third-party audits and internal audits of upcoming periods due to a more transparent information storage system. Further use of the system will allow the organization to function effectively and maintain competitiveness in the market by reducing resource losses.

Acknowledgements The research is carried out Tomsk Polytechnic University within the framework of Tomsk Polytechnic University Competitiveness Enhancement Program grant, Siberian State Industrial University and Kazan State Power Engineering University.

References

1. Surzhikov, A.P., Pritulov, A.M., Lysenko, E.N., Sokolovskiy, A.N., Vlasov, V.A., Vasendina, E.A.: Calorimetric investigation of radiation-thermal synthesized lithium pentaferrite. J. Therm. Anal. Calorim. **101**(1), 11–13 (2010)
2. Kagermann, H., Wahlster, W., Helbig, J.: Securing the future of German manufacturing industry: recommendations for implementing the strategic initiative industrie 4.0. Final Rep Ind **40**, 1–84 (2013)
3. Schuelke-Leech, B.-A.: A model for understanding the orders of magnitude of disruptive technologies. Technol. Forecast. Soc. Change **129**, 261–274 (2018)
4. Lysenko, E.N., Malyshev, A.V., Vlasov, V.A., Nikolaev, E.V., Surzhikov, A.P.: Microstructure and thermal analysis of lithium ferrite pre-milled in a high-energy ball mill. J. Therm. Anal. Calorim. **134**(1), 127–133 (2018)
5. Lysenko, E., Nikolaev, E., Vlasov, V., Surzhikov, A.: Microstructure and reactivity of Fe_2O_3–Li_2CO_3–ZnO ferrite system ball-milled in a planetary mill. Thermochim. Acta **664**, 100–107 (2018)
6. Camiña, E., Díaz-Chao, Á., Torrent-Sellens, J.: Automation technologies: long-term effects for Spanish industrial firms. Technol. Forecast. Soc. Chang. **151**, 119828 (2020). https://doi.org/10.1016/j.techfore.2019.119828
7. Habrat, D.: Legal challenges of digitalization and automation in the context of industry 4.0. Procedia Manuf. **51**, 938–942 (2020). https://doi.org/10.1016/j.promfg.2020.10.132
8. Jayasekara, D., Lai, N.Y.G., Wong, K.-H., Pawar, K., Zhu, Y.: Level of automation (LOA) in aerospace composite manufacturing: present status and future directions towards industry 4.0. J. Manuf. Syst. **62**, 44–61 (2022). https://doi.org/10.1016/j.jmsy.2021.10.015
9. Tussyadiah, I.P., Tuomi, A., Ling, E.C., Miller, G., Lee, G.: Drivers of organizational adoption of automation. Ann. Tour. Res. **103308**, 0160–7383 (2021). https://doi.org/10.1016/j.annals.2021.103308
10. Sapargaliyev, D., Shulenbayeva, K.: Informatization of Kazakhstani higher education. Procedia. Soc. Behav. Sci. **83**, 468–472 (2013). https://doi.org/10.1016/j.sbspro.2013.06.092
11. Obukhov, S.G., Plotnikov, I.A., Surzhikova, O.A., Savkin, K.D.: Method for prediction of the power output from photovoltaic power plant under actual operating conditions. IOP Conf. Ser.: Mater. Sci. Eng. **189**(1), 012008 (2017)
12. Treber, S., Breig, R., Kentner, M., Häfner, B., Lanza, G.: Information exchange in global production networks: increasing transparency by simulation. Stat. Exp. Sel. Digit. Act. Procedia CIRP **84**, 225–230 (2019). https://doi.org/10.1016/j.procir.2019.04.214
13. Vermaelen, N., Kovach, J.V.: Driving meeting effectiveness through organizational process improvement—A Lean Six Sigma case study. Organ. Dyn. **51**(2), 100827 (2022). https://doi.org/10.1016/j.orgdyn.2021.100827
14. Amankwah-Amoah, J., Khan, Z., Wood, G., Knight, G.: COVID-19 and digitalization: the great acceleration. J. Bus. Res. **136**, 602–611 (2021). https://doi.org/10.1016/j.jbusres.2021.08.011
15. ISO 19011:2018 Guidelines for auditing management systems. https://www.iso.org/standard/70017.html

16. Graf, S., Gérard, S., Haugen, Ø., Ober, I., Selic, B.: Modelling and analysis of real time and embedded systems. In: MODELS 2006: Models in Software Engineering, pp. 126–130 (2006). https://doi.org/10.1007/978-3-540-69489-2_16
17. Bolek, V., Filanová, J., Ondrášová, I., Martinková, J.: Process modeling of internal audit in healthcare center. Kontakt **17**(3), 154–162 (2015). https://doi.org/10.1016/j.kontakt.2015.08.006

The manufacturer's authorised representative in the EU is Springer Nature Customer Service Centre GmbH, Europaplatz 3, 69115 Heidelberg, Germany. If you have any concerns regarding our products, please contact ProductSafety@springernature.com

Printed and bound by CPI Group (UK) Ltd, Croydon, CR0 4YY
01/10/2026
02225919-0006